The Unity Principle

The Unity Principle

The Link between Science and Spirituality

Steven L. Richheimer, Ph.D.

Second Edition

InnerWorld Publications
San Germán, Puerto Rico
www.innerworldpublications.com

Copyright 2011, 2013 by Steven L. Richheimer

All rights reserved under International and Pan-American Copyright Conventions.

Published in the United States by InnerWorld Publications, P.O. Box 1613, San Germán, Puerto Rico, 00683

Library of Congress Control Number: 2013920191

ISBN 9781881717294

Cover Design: Rodrigo Adolfo

All rights reserved. This book, or parts thereof, may not be reproduced in any form or by any means, electronic or mechanical, including photocopying, recording, or by any information storage or retrieval system, without permission of the publisher except for brief quotations.

Dedication

This book is dedicated to P. R. Sarkar (Baba/Shrii Shrii Anandamurtiji), who has been my guru for the past four decades and inspired me to write this book. He was a personality incomparable to any other and undoubtedly the greatest spiritual teacher of the twentieth century. He has been an inspiration to me and to millions of his followers.

Acknowledgments

I would like to thank my wife and spiritual companion, Jeanne P. Richheimer, for her help in editing this book and for helping with the illustrations. I am also deeply indebted to Devashish Donald Acosta for his help in editing and laying out the final manuscript for this publication.

Contents

Introduction	1
The Need for a Different Terminology	6
1. The New Physics of Quanta	7
Indeterminacy and Complementarity of Quantum Mechanics	8
The Nonlocality of Quantum Physics	9
The Experimental Proof for the Nonlocality of Quantum Particles	12
The Twin-photon Experiment of Nicolas Gisin	15
The Holistic Explanation of Quantum Physics	15
2. Time, Space, and Relativity	17
How Time and Space Began	17
Problems with the Big Bang Model	18
Light Speed, Relativity, and Einstein's Theories	22
The Modern Concept of Time	26
The Meaning of Now	27
Nonlocality of Time	28
The Arrow of Time	29
Time Travel	30
Consciousness and Time	32
Time According to the Metaphysics of Spiritual Philosophy	33
3. Simple is Better	38
Ockham's Razor	38
Consciousness is a Product of Matter	39
Nonlocality and Quantum Connectivity	39
Psychic Abilities	40
Homing and Other Animal Instincts and Behavior	42
The Goldilocks Enigma	44
A Theory of Everything	47
Evidence for Conscious Design	48
The Universal Mind of Man	49
4. Brahma Chakra: The Cycle of Creation	51
Monistic Religions and Spiritual Traditions	51

Brahma the Cosmic Entity — 52
The Qualification of Brahma — 53
The Dominance of Static *Prakriti* — 53
The *Prati-saincara* Phase of Creation — 57
Development of Intellect — 61
Evolution of Man — 63
Spiritual Practice or Sadhana — 64
Path of Bliss: The Final Journey of Man — 65
The Purpose of Life — 66
Other Theories of Creation — 67

5. The Rule of Three — 68

The Three Forces of the Physical Realm — 68
The Three Principal Layers of Mind — 73
The Three Phases of Life — 74
Food — 77
Everything Else — 78

6. The Law of Action and Reaction — 79

Sadhana and Samskaras — 81
Samskaras and Death — 82

7. The Nature of Unit Consciousness and Unit Mind — 84

The Layers of Unit Mind — 85

8. Happiness, Suffering, Good, and Evil — 90

9. Life after Death — 99

Remembering Past Lives — 100
Near-Death Experiences — 101
Belief in Reincarnation — 102
Paranormal Beings — 103

10. Psychic Body and Power — 106

The Chakras — 107
Psychic Powers and ESP — 110
Occult Powers — 112

11. Meditation — 114

Tantra — 115
Morality and Dharma — 115
Meditation Practices — 116

12. The Unity Principle in the Teachings of Religious Prophets — 119

    Sadashiva (~5000 BCE) — 119
    Lord Krishna (~1500 BCE) — 120
    Moses (~1400 BCE) — 121
    Confucius (~550 BCE) — 122
    Lao Tzu (~500 BCE) — 122
    Gautama Buddha (~400 BCE) — 123
    Patanjali (~150 BCE) — 125
    Jesus of Nazareth (~4 BCE-29 AD) — 125
    Muhammad (570-632 AD) — 127
    Shankara (~790-820 AD) — 127
    Shrii Shrii Anandamurti (1922-1990) — 128
    Other Contemporary Personalities — 129

13. The Unity Principle in Scripture — 131

    Upanishads — 131
    Bhagavad Gita — 132
    Taoist Scriptures — 133
    Ananda Sutram — 134

14. Scientists on Unity — 135

    Albert Einstein — 135
    Max Planck — 136
    Wolfgang Pauli — 137
    James Jeans — 138
    Erwin Schrödinger — 140
    Werner Heisenberg — 142
    David Bohm — 143
    Henry Margenau — 144
    Fritjof Capra — 146

15. Western Attempts to Understand Unity — 149

    The Old School: God is Separate from His Creation — 149
    Panentheistic Concepts of God and Creation — 150

16. The Mystical Vision of Unity — 153

    What is Mysticism? — 153
    Mystics throughout the Ages — 154
    St. Catherine of Siena (1347-1380) — 155
    Ramakrishna (1836-1886) — 156
    Robert Adams (1928-1997) — 157

Gopi Krishna (1903-1984)	158
Eckhart Tolle (1948-present)	158
Author's Experience	159

17. Unity — The Better Explanation — 162

The Multiple Problems with Material Realism	162
The Better Explanation for Reality—The Unity Principle	166

18. Practicing Unity — 171

The Ego Problem	171
Unity and the Human Condition	174
Traditional Practices for Experiencing Unity	177
Surrender, Service, and *Madhuvidya*	182
Meditation, the Most Effective Practice	184
Devotion: The Final Stage	185

Glossary — 187

Notes — 199

Index — 207

Introduction

Material realism (MR) is also called materialism or scientific materialism. It is the theory that physical matter, along with energy, is the only reality and that all phenomena, including the mind, can be explained in terms of the physical interaction of matter and energy. MR takes the position that everything in the universe can ultimately be known through the senses and by physical measurements.

The theory of MR was an outgrowth of the physics of Newton and other eighteenth- and nineteenth-century scientists. The mechanistic Newtonian physics advanced our understanding of the motions of celestial bodies and physical objects. During the same period, Darwin provided a reasonable explanation for the evolution of complex living organisms. Nineteenth-century scientists began to see the universe as a finely tuned mechanistic system, much like a clock, and believed that they were on the verge of explaining everything in terms of matter, energy, and physical forces. There was no longer a need to draw upon a supernatural being or God to explain the workings of the universe.

MR claims that mind and consciousness originate from matter and energy; that mind and consciousness have no existence outside the material brain or nervous system. Secondly, MR states that mind or consciousness can have no direct effect upon physical objects—strong objectivity. It also assumes that all phenomena can be predicted exactly if the initial conditions are known and the laws of physics are applied—causal determinism. The fourth assumption is that objects, no matter what size or mass, are governed only by local force fields and constrained by the speed of light, which cannot be exceeded—locality of matter-energy interactions. Finally, material realism says that life resulted from random chemical permutations and that more complex living organisms and structures result from random mutations and selection of the fittest—material Darwinism.

What exactly is wrong with the theory of material realism? It turns out that many scientifically observed facts and observations contradict the postulates of MR. Science teaches that if a theory fails to account for observed physical phenomena then that theory must be modified or thrown out. In fact, if even one of the six assumptions of MR were proven false then the validity of the theory as an explanation for reality would need to be rejected. As we will see in the first part of this book, modern science has proven that just about all the assumptions of MR are either false or do not stand up to scientific observations.

Let us start with the first assumption—mind and consciousness originate from matter. This assumption implies that the faculties of mind and awareness originate from crude matter and energy. This assumes that the subtlest and apparently most nonlocalized faculty, consciousness, originates from matter. Logic might suggest the opposite—that mind and physical matter arise from consciousness. That is, that consciousness is the starting point for creation and that it is first transformed into mind and then into matter.

Furthermore, if consciousness were solely a function of electrochemical brain activity then we might expect that there would be some physical structure in the brain responsible for it. However, scientists and neurosurgeons have not been able to isolate or identify a part of the brain or neurotransmitter that is responsible for consciousness. Large portions of the brain can be lost to disease, injury, or surgery, but consciousness and self-awareness, albeit diminished in many cases, remain intact. Hence, mind and consciousness seem to pervade the entire brain in both humans and animals. Although one may argue that a functioning brain is required for the expression of consciousness in living organisms, the same cannot be said for the converse—since consciousness appears to function outside the boundaries of the physical body.

Material realism denies the possibility that mind and consciousness can function outside the brain or physical body. Thus, all out-of-body phenomena are assumed false or illusionary. Such phenomena include:

ESP (extrasensory perception)
Life after death
Near-death experiences
Mind (astral) travel
Mystical and religious experience

Dean Radin's book, *The Conscious Universe* helps to dispel any doubts about the scientific validity of ESP.[1] Statistical meta-analyses of hundreds of well-controlled scientific studies provides unquestionable scientific evidence for the factual existence of ESP capabilities in human beings. ESP studies show that the human mind is capable of accessing information nonlocally in time and space. One of the best examples of this is remote viewing in which the trained or gifted viewer is able to describe in detail a scene witnessed by another person—even before that person arrives at the target location. This capability has been extensively studied under controlled laboratory conditions and even used by espionage agencies in the past to gather intelligence information. The factual existence of psychic phenomena point to a universe in which information can be passed to an individual nonlocally. The factual existence of ESP alone disproves the myth of materialism.

Survival of consciousness after death is a postulate of all the world's religions. Survival of consciousness after death is difficult to prove, but numerous reports of out-of-body experiences during near-death or during unconsciousness, such as surgical procedures, point to the ability of the mind to function outside the body. Such reports would not be considered to have scientific validity if it were not for many subjects' uncanny ability to describe events, features, and dialogue that occurred while that person was unconscious.

There have been many reports and studies indicating that some children have accurate memory of events, names, and locations of a previous life on earth, and there are many reports of hypnotic regression bringing out specific memories of a previous life on earth. Most often regression is performed on people suffering from a neurosis or irrational fear. Under hypnosis, there is recollection of a traumatic event experienced in a previous body, and the resulting catharsis yields an unexpected cure for the neurosis.[2]

The willful act of separating the mind from the physical body is sometimes called astral projection. Unlike an out-of-body experience, which happens involuntarily, astral projection is done by persons who have received instruction and have trained for this experience. There has been little scientific study of astral projection; however, people adept at remote viewing may be practicing a form of mind traveling.

Throughout the ages, there have been reports of prophets, sages, saints, rishis, and mystics that have experienced connecting with the cosmos. This experience has been described in various ways such as mystical experience, Cosmic Consciousness, enlightenment, liberation, self-realization,

God-consciousness, samadhi, moksha, nirvana, etc. Since these are individual experiences, there is no way of scientifically testing their validity. It is apparent, however, that people are truly changed by these experiences and that most of the world's religions, spiritual movements, and cults have been spawned by mystics.

As we will see in Chapter 1, the assumption that mind and/or consciousness can have no direct effect upon physical objects (strong objectivity) is contradicted by the experimental observations of quantum mechanics. Countless experiments have shown that quantum events are not observer independent. The simple act of observing or measuring a quantum system will change the system. At best, one could say, as Bernard d'Espagnat suggests, that the objectivity of quantum mechanics is weak objectivity, that is, that the outcome of the observation of a quantum event is not dependent on who makes the observation.[3] The experiments indicate that physical reality is changed by the intervention of consciousness. The falsehood of strong objectivity is another reason to reject the hypothesis of MR.

The assumption of causal determinism and locality of physical interactions are at the heart of the materialist explanation of reality. Strong causal determinism assumes that if all the initial conditions of a system are known then any change in the system caused by an outside agent (whose condition is also known) can be predicted precisely. Locality is the principle that all interactions between particles occur by local interactions or fields through space-time. Observations of quantum particles and phenomena contradict these assumptions. For example, the randomness and unpredictability of radioactive decay contradicts the assumption of material realism that the state of a system can be predicted exactly if initial conditions are known. In fact, all quantum systems have a small degree of uncertainty, which means that it is never possible to describe the system exactly. Causal determinism does not apply at the fundamental level of matter and energy—only probabilities apply.

In Chapter 1, we will also see how quantum observations of the nonlocality of time and space contradict the assumption of locality of physical interactions. Hence, modern scientific observations contradict the theory of strong causal determinism and locality of physical objects.

We will see how the empirically derived data of quantum physics and relativity theory demonstrate that on the tiniest (quantum) level and on the largest macroscopic (cosmic) level, the universe seems to behave as an undivided Whole. This is precisely the same conclusion that mystics, seers, and prophets arrived at based on their subjective experience of reality. A universe that is One or a Singularity implies that whatsoever exists in the microcosm must exist in the macrocosm and vice versa. Since human beings possess consciousness then it follows that the macrocosm possesses consciousness and that consciousness is the true building block of the universe—not matter.

The undivided wholeness of the created universe is essentially what we call the Unity Principle. To many in the West it is a foreign concept that goes against everyday experience, for it implies that discreetness, differentiation, individualism, etc. are merely a macroscopic illusion. However, to those schooled in one of the spiritual traditions of the East such as Vedanta, Buddhism, Tantra, yoga, Taoism, or Sufism this Principle is at the core of their understanding of reality. These philosophies and religions of holism are also cyclical in nature: creation begins and ends with Consciousness.

The Unity Principle also goes by the terms: monism, monistic idealism, holism, and monistic panentheism. Unlike the philosophy of dualism, which teaches that the universe consists of, or is explicable as two or more fundamental entities, such as matter and mind, living and nonliving, God and creation, etc., monism purports that the universe is a Singularity (One). Everything is connected, and all things originate from Supreme Consciousness.

Science is mainly concerned with obtaining experimentally derived information about physical reality. Not surprisingly, most scientists subscribe to the idea that this physical reality is the ultimate reality, i.e. material realism. Today scientists display a great prejudice toward material realism and scientific objectivity, even though MR has been repeated shown to be false by experimental observations. Why are scientists reluctant to bury the corpse of material realism and allow consciousness to take its rightful place in science? Perhaps it is because to do so would be to open the scientific realm to metaphysics and allow an idealist revolution to sweep away the obsolete doctrine of material realism.

The problem is that material realism is a dualistic theory that tends to degrade the human mind by assigning a firm reality to the physical world. However, when experimental observations contradict the assumption that matter is the ultimate reality there is a need to reject the old philosophy of material realism and find a new model to describe reality. The simplest and most logical model that is consistent with the empirical evidence is the Unity Principle, and more and more scientists have come to embrace this new paradigm of reality.

The Unity Principle is a theory of everything. It provides an explanation of how the material universe is created from Consciousness, and it goes on to explain how matter can be transformed into living organisms, and how and why life forms evolve toward beings that are self-aware and thirst for the Infinite. We will discover how the Unity Principle unveils many of the mysteries of science, nature, religion, spirituality, and the purpose and meaning of life.

The Need for a Different Terminology

The explanation of how Consciousness is transformed into the material world is best done using some Sanskrit words since the English language does not adequately express many of the concepts discussed in this book. Whenever possible the best English equivalent to the Sanskrit word is given. The reader may also take advantage of the extensive glossary at the end of the book.

1
The New Physics of Quanta

By the early twentieth century, the "classical physics" of Newton was being replaced with quantum mechanics and the relative mechanics of Einstein. Although quantum mechanics is successful in describing the behavior of microscopic particles, it has limitations on the macroscopic scale. On the other hand, Einstein's relativistic mechanics describes the gravitational forces and motion of large bodies in space-time dimensions, but does not address the interactions of small particles.

Because of the incompatibility of these two theories, scientists have been seeking a "theory of everything" that would unify quantum and relativistic mechanics in one grand theory. Today the closest physicists have come to this goal is string theory. This theory mathematically describes all particles as vibrating strings and utilizes extremely complicated mathematics and multi-dimensional analyses. So far, it has failed in its attempt to be a unifying theory and physicists doubt that they will ever be able to test any of its hypotheses. Physicists have failed to find a unified theory of everything because they doggedly stick to the misconceived paradigm of a dualistic universe. They have found it necessary to invent complicated and unreasonable theories such as many-worlds and multidimensional analyses to explain the observed nonlocality of quantum particles. Scientists may well have to discard the materialistic model of the universe and come to an understanding of the unity of all things before they can explain in a simple and logical way the true nature of reality.

Quantum physics strongly suggests that everything in the universe is connected. However, to understand why this is the case we should first explore some of its ideas regarding reality beginning with two of its most fundamental concepts: indeterminacy and complementarity.

Indeterminacy and Complementarity of Quantum Mechanics

Indeterminacy simply means uncertainty. It is one of the cornerstones of quantum mechanics. First described by Werner Heisenberg, his famous uncertainty principle stated that the uncertainties in measuring the position and the momentum of a quantum particle are always equal to or greater than a constant (Planck's constant). That is, the greater the precision with which the position of a quantum particle such as an electron is measured, the less precisely is its speed known, and vice versa. Inherent in this principle is the observed phenomenon that the very act of measuring the position of the particle affects it, and hence, causes uncertainty in measuring its speed. We can ignore the fuzziness, which is inherent in the quantum realm and relativistic effects of motion in everyday life because they are so small. Because of this, the classical descriptions of nineteenth-century physics normally work well on the macro scale.

Complementarity in the quantum realm describes the entangled and ambivalent qualities of quantum particles or phenomena. For example, the dual aspects of light as both wave and particle are complementary. Neither description for light works under all experimental conditions. A complete description of light requires that we consider both of these mutually exclusive constructs, and our knowledge of the situation is necessarily limited because we are unable to simultaneously measure or describe both constructs precisely (uncertainty principle).

The existence of complementary aspects for describing a physical construct indicates that the true reality of the construct is to be found at a deeper level than that of the two complementary descriptions. Examples of complementarity in the physical realm are: wave-particle, space-time, position-momentum, and matter-energy. Today complementarity is the logical framework for comprehending and acquiring scientific knowledge in the physical realm. The concept of complementarity can also be applied to other fields besides quantum mechanics because there is no single or simple way to describe reality from our limited perspective. Examples of this type of complementarity are: particular-contingent, cause-effect, observed-observer, thought-action, physical-psychical, yin-yang, male-female, and part-whole.

One of the weirder aspects of quantum mechanics is that on the quantum level something can simultaneously exist and not exist. For example, if a subatomic particle is capable of existing in several different states the uncertainty principle allows it to exist in all possible states at the same time. When the particle is measured or observed by physical means and its state is no longer uncertain, the act of measurement instantly forces it into just one state. This is called the "collapse of the wave function." For some particles that are paired, the observation of one particle causes the wave function of both particles to collapse into the same or opposite state even if the particles are separated by great distance.

The Nonlocality of Quantum Physics

The study of quantum physics leads inexorably to the conclusion that the quantum world exists in both temporal and spatial nonlocality. What is meant by temporal nonlocality? Simply put it implies that the past, present and future coexist in the now and that time does not flow linearly from the past, to the present, to the future. That is, effect can and many times does precede cause.

Spatial nonlocality means that quantum particles cannot be placed with certainty at any given set of coordinates. Only a probability function can define their location and there is always a finite probability that they can be found anywhere in the universe. Another way of looking at this is to consider that quantum particles have wave nature that can be spread out through space and time.

One of the classic experiments of modern quantum physics that reveals the nonlocality of space and time is the dual slit experiment. In this experiment, light (photons) or electrons are passed through two slits that are very close together. Both electrons and photons have both wave and particle characteristics depending on how they are observed. Typically, the wave-particles passing through the slits interact as waves to form an interference pattern (alternating dark and light lines) as the waves either cancel or reinforce one another. It is not difficult to understand the formation of this interference pattern when many particles are passing through the slits simultaneously and act as waves.

Interestingly if the flow of particles is slowed so that only one photon passed through the apparatus at a time the pattern persists. Obviously, the individual photons cannot split, each half going through a different slit and then recombine before they hit the detector. Therefore, the interference pattern should disappear since it is created by the interaction of wave fronts from the two slits. By exposing a photographic plate for many minutes or summing the response from an electron detector, scientists observe an interference pattern suggesting that the so-called "well-defined" particles are behaving like a wave or individual particles are interacting with particles that passed through the slit at a different time. That is, they exhibit either spatial or temporal nonlocality.

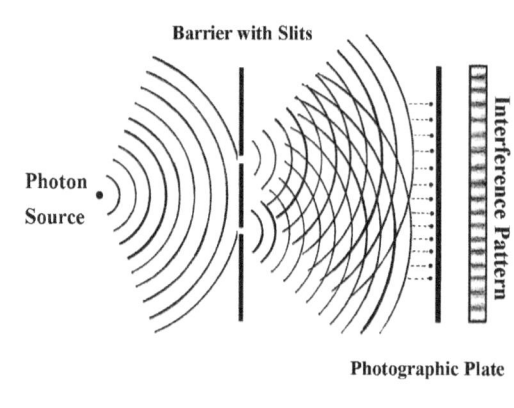

Dual Slit experiment. Waves from the photon (light) source pass through two closely spaced slits causing the wave fronts to interfere with each other producing alternating dark and light lines.

In another experiment, an apparatus is used that has a very thin coating of silver that creates a 50 percent chance that a photon will pass through and a 50 percent chance it will be reflected.

If the photon passes directly through the half-silvered mirror, it takes a direct path through the slits onto the photographic plate, while if it is reflected it must take a longer route and arrive at the photographic plate later. It should be an either/or situation and there is no reason to expect the photon to interfere with itself and create an interference pattern. However, development of the photographic plate reveals an interference pattern, which indicates that the photon is somehow enmeshed with itself across time. This is a classic example of temporal nonlocality at the quantum level.

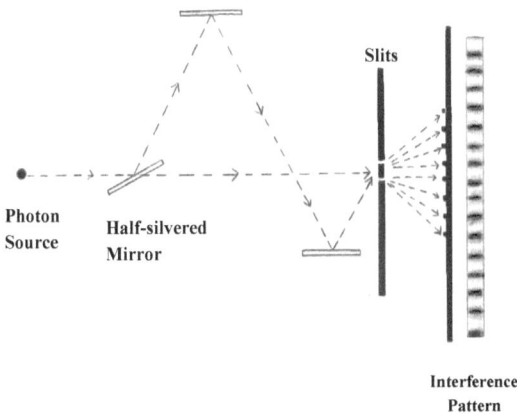

Modified Dual Slit experiment demonstrating temporal nonlocality. Depending on which path it takes, a photon will pass through the slits and arrive at the photographic plate at different times.

The dual slit experiment can be modified in such a way that one of the slits is closed electronically while the photon is in flight toward the apparatus or after it has passed through a slit but before it strikes the detector. In both cases, the interference pattern disappears and only a simple diffraction pattern is observed. Theoretically, if the photon is forced to pass through only one slit it is forbidden from acting like a wave and reverts to its particle nature. The same is true when the slit is closed after the photon passes through one of the slits, except that this effect precedes the cause (temporal nonlocality). However, the question is, how does the photon that goes through the open slit know that the other slit is closed or about to be closed after it passes through and that it must go to a different location on the photographic plate? Somehow, the photon knows the other slit is not fully available to it and it acts accordingly. Does this mean that the photon behaves in a conscious manner? Perhaps so. In any case, such experiments clearly demonstrate that the behavior of the particle is not just determined by the conditions of the test. The wavelike or particle-like behavior is determined by the decision of the experimenters as well as by the observation they make. This outlines an important aspect of quantum nonlocality, namely, that the observer and the observed system cannot be separated. The observer or his instruments are part of the system and influence the outcome of the observation. In other words, the act of observing alters or influences the system and this alteration occurs outside linear time. The observation can occur after the particle has passed through the slits, yet the way the system is observed still influences the outcome or behavior of the system. This is contrary to the common notion of cause and effect since on the quantum level the effect can precede the cause.

The Experimental Proof for the Nonlocality of Quantum Particles

Quantum mechanics was first introduced by Neils Bohr and his collaborators in Denmark in the early part of the twentieth century. The theory required that quantum particles be nonlocal in nature. That is, we can never know with certainty both the position and momentum of a quantum particle; at best we can only assign a probability that it will be found at a particular location.

More importantly, when we have two or more paired quantum particles they do not behave as separate entities. Quantum theory required that, for example, when the spin of one of two correlated quantum objects is measured it affects its partner, even if the two objects are on opposite sides of the galaxy. The particles are said to be entangled or enmeshed. The rules governing the theory required that when one particle in a pair changes one of its properties the other paired particle must change to an opposite state simultaneously.

However, this nonlocality of quantum objects went against the material realism bias of Einstein and his colleagues. They argued that according to quantum theory, quantum particles separated in space would have to interact with one another instantaneously via local signals at a speed greater than that of light. Einstein called this "spooky action at a distance." Einstein felt that signals between particles could never exceed the speed of light and if quantum mechanics predicted faster than the speed of light communication between particles then there had to be a serious flaw with the theory.

The real issue here was separability. The material realism of Einstein considered all objects as separate entities, while the new physics of Bohr required that correlated quantum particles were nonlocal and hence connected through space-time. Einstein and Bohr, along with other scientists, had a running feud about the nonlocality of the quantum realm that lasted nearly twenty-five years but was finally resolved by experiments performed after the death of both men.

In 1964 physicist, John Bell proposed an experiment that would conclusively prove which version of reality was correct. Today it is called Bell's theorem, but it took another 20 years before the experiments could be conducted which would demonstrate conclusively that nonlocality was a fact of nature.

The definite experiment testing Bell's theorem was conducted by Alain Aspect and his colleagues at the University of Paris-Sud in 1981. Aspect set up an apparatus that emitted paired photons. That is, the two photons were emitted simultaneously from an atom with the same quantum numbers except for their polarization. These are in a singlet state, or correlated quantum particles. Born of the same source with wavelike aspect, they are never truly separate but are entangled, two parts of the whole. According to quantum mechanics, these particles are intimately connected, and it is not necessary for any signal or field to pass between the particles for one to know what the other is doing. Hence, if the polarization of one is changed or measured, quantum theory requires that the polarization of the other particle must change simultaneously in such a way that the polarization of the two particles always remains opposite.

Polarization is simply the angle at which the electrical vibration occurs relative to a plane for electromagnetic energy such as light. Most people are familiar with polarized sunglasses that block horizontally reflected sunlight, reducing glare and making it much easier for anglers to see fish in the water. The detectors that Aspect used were similar to such glasses except that they could be turned to any angle at will. Quantum mechanics requires that the polarizations of the two correlated photons are indeterminate until one is measured, but when this occurs then the other photon must have an opposite angle.

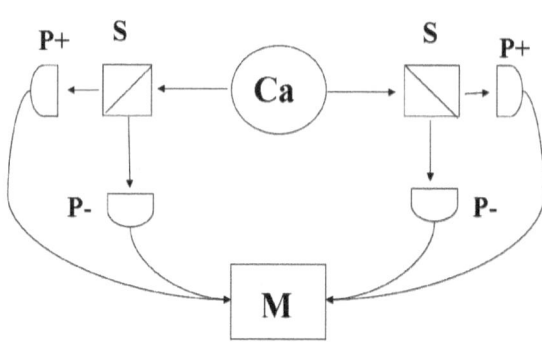

Diagram of the Aspect Experiment. The calcium ion source (Ca) produces entangled or paired photons sent in opposite directions. Each photon encounters a fast optical switch (S) and a two-channel polarizer (P) whose orientation can be set by the experimenter. Emerging signals from each channel are detected and coincidences counted by the coincidence monitor (M).

As predicted by quantum theory the two photons were always found to have opposite polarizations. This observation alone did not prove nonlocality since it is possible that the two photons could be created with opposite polarizations, just like a left and right hand and always be different. Alternatively, they might communicate via local fields with one another so that when the wave function of one was collapsed by observation it told the other photon to collapse in the opposite state. However, Aspect added an electronic polarization switch that had the effect of changing the polarization setting of one of the detectors every ten-billionth of a second—shorter than the time for light to travel between the two detectors and a test using Bell's theorem eliminated the possibility that the particles were created with opposite polarizations that never changed. Clearly, the change of the polarization setting in one detector influenced the outcome of the measurement in the second detector showing that the two photons were indeed connected through space just as predicted by quantum theory. There was simply no time for the information about the change in one detector to be transmitted to the other paired photon through local signals.

The Twin-photon Experiment of Nicolas Gisin

In 1997 Nicolas Gisin of the University of Geneva and his colleagues sent linked pairs of photons along fiber optic cables in opposite directions to villages north and south of Geneva, Switzerland.[1] According to classical physics there would be no way these photons could communicate with each other. However, when the independent polarization decision of one photon was compared to its twin their polarizations were always opposite, even though there was no way for them to communicate with each other. As in the Aspect experiment, the photons were entangled. They shared a common origin and properties and remained in instantaneous touch with each other, no matter how far apart they were. In a sense, entangled quantum particles are not discrete separate entities. They have wave-like properties that allow them to spread throughout the universe and touch one another. They are never truly separate in the first place.

The Holistic Explanation of Quantum Physics

Today the nonlocality of quantum fields and particles (entanglement) is an accepted fact of nature that has been observed for not only photons but also electrons, atoms and even large molecules. It is the basis for the development of today's quantum computers, which one day might revolutionize electronics. Yet the connection that exists between distant but entangled particles is one of the greatest mysteries of quantum mechanics. Entanglement of particles is not constrained by either time or space. Either a field of information pervades all of space-time, which allows quanta to communicate with one another, or all quanta must be connected and none is truly separate from another. Such an explanation for the phenomenon of quantum nonlocality is consistent with the idea that oneness pervades all space and time. However, today most physicists do not accept this idea, something we call the Unity Principle, since it touches on the metaphysics of holism.

Our current understanding of the origin of the universe suggests that it began as a single point or singularity containing all the mass-energy of the universe and has been evolving ever since what scientists have dubbed the "Big Bang." However, if all quanta interact with one another or are entangled, they, and the entire universe can be thought of as being merely parts of a quantum singularity, parts of the Whole. Nonlocality in the quantum realm implies that by nature the universe is nonlocal and can best be understood as an indivisible Whole.

Furthermore, experiments probing the quantum realm consistently show that the experiment is influenced by the intent of the experimenter. In other words, the experimental system, which includes the mind of the experimenter, comprises an unanalyzable whole.

Hence, modern physics proclaims that the constituents of matter-energy and all phenomena involving them are intimately connected and interdependent. Such phenomena can best be understood as parts of the Whole, which includes the consciousness or mind of the observer.[2] According to this idea the discreteness or individuality we observe in the objective universe must be considered a macroscopic illusion—all things are actually inseparable parts of the One.

This vision of the universe as a unified Whole, the Unity Principle, is the new metaphysics of holism. Quantum physics reinforces this view of the universe, as do many other scientific and subjective observations.

The metaphysics of holism must also apply to living beings. Although Rene Descartes had it backward when he famously stated, "I think therefore I am," it is nonetheless true that we possess awareness or consciousness. If we possess consciousness and the universe consists of a unified Whole, then the cosmos possesses consciousness. Holism requires that what exists in the microcosm must also exist in the macrocosm. Since consciousness is a subtler faculty than any subatomic particle or building block of the physical universe, it follows that everything in the material world has evolved from Consciousness. Similarly, it follows that living organisms and mind have also evolved from Consciousness. Therefore, Consciousness may be understood as the great unifying principle that makes the metaphysics of holism possible.

2

Time, Space, and Relativity

Thanks mainly to discoveries related to the theories of Einstein, today we know that time and space are complementary aspects of the same thing, what scientists call space-time. Hence, time and space are not independent but fully entwined with one another in four-dimensional space-time. Intuitively it is difficult to understand how space and time are interconnected and dependent on one another. However, it should be obvious that the existence of time depends on space, for if there is no space then there is no need for the concept of time that would be required to traverse it. In order to understand the modern concept of time and space we should first review what cosmologists know about how the cosmos originated.

How Time and Space Began

Cosmologists believe that time and space began in a gigantic explosion that occurred some 13.7 billion years ago in what is commonly called the Big Bang. The principal reason for this theory is that the universe is known to be expanding. Discovered by Edwin Hubble, distant galaxies are all moving away from Earth and from each other at a rate proportional to the distance separating them from each other and that the rate increases with time. This is called Hubble's law and the rate of expansion is called the "expansion factor" or "scale factor." Hence, if a galaxy is one hundred light years distant its velocity might be V. However, for a galaxy that is two hundred light years distant it is found to be receding from Earth at 2V. This expansion is not due to the mechanical results of the Big Bang but to the expansion of space itself. If this were not so then distant galaxies would be moving away from us at a speed greater than light, which is not possible.

Another consequence of this type of expansion is that the universe, which began as a point, has no center. No matter where you stand in the universe, everything is observed to be receding from you at the same relative rate depending on its distance from you. Therefore, the universe has no center or the center can be considered everywhere.

Prior to the Big Bang space-time did not exist; therefore, it is meaningless for scientists to try to hypothesize about what came before the Big Bang.

The linear concept of space-time plays an important role in the physical sciences such as geology, astronomy, physics, and biology, but linear time is a classical concept. In both relativistic and quantum physics as well as in metaphysics the linear concept of time is meaningless.

Problems with the Big Bang Model

The expanding universe model requires that space expand at a certain rate since the time of the Big Bang. However, this model has several problems built into it. One problem is the size of the universe. If the universe began as a point and first expanded at the speed of light then after one year its size would be one light year (the distance light can travel in one year). After approximately 14 billion years, its size should be 14 billion light years. However, the observed size of the universe is some 46 billion light years. How is this possible? In addition, the universe is known to be extremely homogeneous, but the mixing that is required for homogeneity could not take place if separate sections of the early universe expanded at or near the speed of light. There would simply be no way they could have contact with one another. To account for homogeneity one must assume that all quanta are entangled and/or the initial expansion of space took place at a speed greater than that of light.

Another difficulty with the model is that of the "flat universe." If the expansion of space is even a tiny bit slower than the counteracting force of gravity then the universe is said to be "closed," gravity would prevail, and the universe would begin to contract leading to a "Big Crunch." The rate of contraction would accelerate rapidly leading to a universe that collapses back to a point in only a few million years. On the other hand, if the rate of expansion of space were slightly larger than the counter balancing pull of gravity, we would have an "open universe." In this case, the rate of expansion would accelerate as mass and hence gravity became diluted and after a few million years, everything in the universe would be so far apart that stars and galaxies could never have formed. There is a very fine line between these two models, but the longevity that results from the flat model is crucial, for it provides the time needed for stars, galaxies, planets, and eventually life to evolve.

Another problem with the Big Bang model is the energy of empty space. This energy is called the "cosmological constant" or "lambda," and it could be either positive (attractive) or negative (repulsive). It represents the gravitational force of space itself. Originally, scientists believed in a static universe that neither expanded nor contracted. This factor was needed in order that the universe not collapse under the force of gravity but remained fixed. However, Hubble's discovery of cosmic expansion disproved the static universe model and at first, there was no need for a fudge factor such as lambda to halt an expanding universe. However, scientists now find that stars that are more distant are receding from us at an accelerating rate, not at all the rate that would be predicted if space was expanding regularly against the force of gravity, which would cause a slight deceleration in the expansion. Lambda, the so-called cosmological constant, is again needed; and clearly empty space has more energy than first believed. This energy of the void is called "dark energy" and it is believed to compose some 75 percent of the mass-energy of the cosmos. Scientists still do not understand what it is.

The dark energy of empty space must have some weird properties, for unlike any other material we find in the universe it is both gravitationally repulsive and not diluted by expansion of space but is always a constant. If this were not so then the universe would quickly collapse once again to a dimensionless point. Apparently, this force acts like a rubber band that increases in tension as space expands and this increase in tension force counteracts the effects of dilution. Since the total curvature of the universe appears to be zero or flat this means that the total mass-energy of the universe, which includes ordinary matter, dark matter, and vacuum energy (dark energy), has to add up to the critical density required for a flat universe. The likelihood of this scenario occurring by chance is extremely unlikely leading to the problem of "fine-tuning" of the universe.

To account for the various anomalies of the Big Bang model of cosmogenesis, cosmologists came up with the idea of inflation. This inflation has nothing to do with your money becoming less valuable, but is the theory that the baby universe expanded unimaginably faster than it does today, and thus its size "inflated." The rate of inflation greatly exceeded the speed of light, but this does not contradict the theory of relativity because space itself expanded and the speed of light was not affected. At the end of inflation, the universe is thought to have decayed into normal matter and energy (radiation) and the temporary hyper-fast expansion gave way to the normal expansion of the universe observed today.

How does this theory solve the problems with homogeneity, size, and flatness of the universe? Theoretically, all the parts of the universe began fully connected in the inflation phase and like completely mixed tea and milk were homogeneous. Hence, the theory of inflation explains why the temperatures and curvatures of different regions of space are so consistent. Importantly, inflation allows physicists to calculate the minute differences in temperature of different regions from quantum fluctuations during the inflationary era, and these quantitative predictions have been confirmed. Because of the very rapid growth of space during the inflation phase, the universe expanded far beyond the size predicted if it was limited to the speed of light.

Inflation also provides a mechanism whereby the universe could collapse initially into a flat state. As we have seen, flatness is both unstable and improbable. Even a slight excess of matter-energy as compared to the power of expanding space (lambda) results in a closed universe that very rapidly contracts to a Big Crunch. On the other hand, if expansion is slightly greater than the gravitational forces then an open universe will result, that leads to emptiness that is unable to support life. The theory is that when ordinary matter finally condensed following inflation the tremendous tension in lambda that was built up during the super-fast expansion suppressed curvature of the universe after which it reverted to its normal strength that we witness today. The theory depends only on the idea that lambda is not diluted by expansion and therefore can overcome the forces that lead to curvature of the universe. Hence, inflation theory explains why the curvature of the universe would be suppressed and could result in the required tuning necessary for a flat universe to emerge following this phase of super-fast expansion.

While inflation theory might explain how our universe emerged with just the right conditions needed for a long-lived universe, it simply substitutes one set of improbabilities for another. The temporary lambda of inflation becomes a normal lambda after inflation and the new value for lambda still needed to be "just right" so that the universe that emerged from inflation had the perfect balance of gravitational energy and mass-energy so that it neither hyper expand or crunch. The fine-tuning problem is not solved since a precise value for normal lambda and the total mass-energy of the universe is still needed in order to account for our long-lived universe that was needed for the evolution of conscious life. In fact, Stephen Hawking calculated that the odds for this occurring by chance are only one part in a million trillion.[1]

In order to explain this enigma cosmologists proposed "multiple universes" or "multiverse" theory. Perhaps our universe has the perfect balance of factors because an almost infinite number of universes were formed after various regions condensed or emerge from inflation—much like the production of a long sheet of bubble wrap. All these universes are separate, do not interact with one another and each collapsed with a slightly different values for the cosmological constants, mass-energy, and physical laws that govern it.

Along with the multiverse theory to explain why our universe seems to be fine-tuned for conscious life one needs to apply the so-called "anthropic principle." This principle states that we would only exist in those universes that were finely tuned for our conscious existence. Thus, while the probability might be extremely small that there is life in most of the universes, this scarcity of life-supporting universes provides an alternative to a conscious design explanation for the origin of the universe.

Turning back the clock on the expanding universe one would witness everything becoming closer and closer together. Eventually a neighboring galaxy that is one hundred light years would be right on top of us and so would a galaxy two hundred light years distant. In fact, all matter in the universe would recede to a single point some 13.7 billion years in the past. However, there is a time in the past when we know that all physical theories of the universe must break down. It is called the "Planck epoch" and it occurs only at the very beginning of the universe, even before inflation at some 10^{-42} seconds following the initial formation of the universe. What took place during this period will always be a complete mystery to science. However, it is postulated that all forces and quanta were unified during this phase of creation. In addition, something wonderful must have taken place during this epoch in order that the conditions for flatness developed, for as we have seen without flatness we would not exist. Lucky for us the conditions set during this initial phase of creation were ideal and it was like winning the lottery not once but ten times in a row.

Light Speed, Relativity, and Einstein's Theories

In 1887 American scientists, Albert Michelson and Edward Morley performed one of the most important experiments ever done. They observed that the speed of light was not affected by whether Earth was moving toward a distant star or away. It contradicted the commonsense notion that speeds should add up. That is, if a bullet were fired from a fast moving car then its speed is greater than if the bullet were fired from a stationary position.

Similarly, you would think that light from a star moving toward Earth at 25 percent of the speed of light should reach Earth quicker than if the star was moving away from Earth at this speed. However, this is not the case. Amazingly, the speed of light is the same in both instances.

Previously, the ideas of Newton postulated that space was unchanging or fixed. This three-dimensional fabric of space always remained constant, and bodies were thought to move within it in precise, predictable ways. Time was also a constant within this rigid structure of space, since all clocks within the universe would have to tick at the same rate. This was the basis for the nineteenth-century concept of a "clockwork universe." Physicist at the time thought they were on the verge of explaining all the workings of the universe in terms of the known laws of physics—many of which were laid down by Newton himself.

However, if the speed of light is always a constant then this idea of space had to be wrong. The Michelson-Morley experiment destroyed the concept of a clockwork universe and proved that neither time nor space could be absolute and unchanging. Einstein realized that something else had to change to account for the constancy of the speed of light for bodies in motion. He sensed that the "something" must be space itself. Space could flex and change, become compressed or expanded according to the relative motion of an object and an observer. The only constant was the speed of light itself or an integrated four-dimensional fabric he called space-time.

These insights led to Einstein's special theory of relativity (STR), which states that the universe has four-dimensions. The three of space—width, length, and height—each integrated with the dimension of time. Time is not a separate dimension needed to locate on object in ever-shifting space-time. Each of the four dimensions of space-time has a spatial and temporal component. This is required from the fact that both space and time are relative to the state of motion.

The faster an object moves, the slower the passage of time, until ultimately at the speed of light, time ceases entirely. In other words, Einstein showed that with motion, space converts to time, and when an object is at rest, time converts to space.

Evidence for these concepts comes from studying unstable subatomic particles that are accelerated near the speed of light in particle accelerators. Such particles decay less rapidly than their brethren that are moving more slowly—exactly as the theory predicts. No object having mass can ever attain the speed of light since Einstein's equations show that the mass of a particle must increase exponentially as it nears the speed of light and become infinite at that speed. This phenomenon has also been shown to occur in particle accelerators precisely as predicted by Einstein.

However, photons carrying electromagnetic radiation such as visible light have no such problem. They have no mass, and travel at the speed of light. Their internal clocks are stopped and hence they do not decay like other particles. At the speed of light, they compress space to a point. In a sense, they do not travel through space as much as transcend or bypass it. For this reason, light does not require any medium for its passage through space. However, photons cannot move through space-time any faster than the speed of light in a vacuum, because although they are outside time they are nonetheless constrained by the compressibility of space. Space can be compressed down to a point, whereupon it disappears, but no further. Not only is it impossible for any physical object to reach the speed of light, it is also impossible for light to exceed that speed.

Einstein's special theory of relativity also postulated the equivalence of mass and energy and gave us the famous equation that energy is equal to mass times the speed of light in a vacuum squared ($E=mc^2$). This equation demonstrates that matter is a condensed form of energy, and to this day physicists express the mass of subatomic particles in terms of energy—normally electron volts.

Einstein's second theory of relativity, termed the general theory of relativity (GTR), describes gravity as a geometric property of space-time. This theory predicts that gravity distorts space and causes light from a distant object such as a star to bend. This prediction has been verified experimentally, as has the existence of black holes—objects with such tremendous gravitational force that nothing can escape their pull, including light.

Several startling and unusual consequences arise from Einstein's new model of the universe. For example, suppose a star explodes or goes supernova in our neighboring Andromeda Galaxy. Even though light travels at the overwhelming speed of 186,000 miles per second, stellar and galactic distances are measured in light-years—the distance light travels in one year. When telescopes on Earth pick up the light from the supernova, astronomers calculate that the star exploded some 2.5 million years ago because that is the time it takes for light to travel the 2.5 million light-years between Andromeda and Earth.

However, the concept of four-dimensional space-time demands that the two *events* (the explosion of the star and witnessing the event on Earth) occur simultaneously. This contradicts our intuitive understanding of the passage of time, but from the standpoint of real or cosmic time, the great distance of space separating us from Andromeda is simply one of the coordinates of space-time and creates the illusion that much time has passed between the events.

Another example of how space and time are intimately connected and inseparable could occur in the distant future as our Sun begins to run out of fuel and expands, threatening to engulf the Earth. Earthlings are forced to move to a habitable planet in another solar system. Such an earth-like world might be the recently discovered planet orbiting the star Gliese 581 that lies twenty light-years distant. Next, assume that technology had advanced significantly in the millions of years before Earth becomes too hot to inhabit. Astronauts in this scenario could travel to this nearby star at 90 percent of the speed of light, so the journey would take approximately twenty-two years.

Since radio communication, which travels at the speed of light, would be impractical, let us assume that the astronauts can communicate instantaneously with scientists back on Earth using telepathy. Traveling at this enormous velocity, the space between the astronauts and Gliese 581 is compressed and the astronauts would calculate their distance to the star as half that calculated by scientists on Earth. However, when the astronauts compared the time that had elapsed since they left Earth with clocks on Earth they would discover that their clocks were running at exactly half the speed as those on Earth. Both sets of measurement are consistent. The astronauts think they will arrive in half the time because they calculate the distance to be half, but because their clocks are slowed down by half, they actually arrive at the exact instant as calculated by scientists on Earth. It would be a boon to the astronauts that the journey took half as much time as expected. Instead of aging twenty-two years, they will age only eleven years during their journey as compared to their counterparts on Earth.

The integration of space and time into a four-dimensional space-time continuum means that time does not pass linearly and only events have any meaning. While the astronauts traveling to a new world cannot agree on how far the journey is or how long it will take, they do agree on the two events—the launch time and the arrival time.

Since all spatial dimensions are compressed at high speed, it means that the shape of their starship would look compressed or flattened to someone on Mars that witnessed the ship as it passed by. Hence, space and time are observer dependent, and time and length may expand or shrink depending on the relative state of motion of the observer and what is observed. As space shrinks, time dilates. Space is transformed into time and time into space. This is the hallmark of a four-dimensional substance in which each dimension has both a spatial and temporal aspect, and they are fully integrated and inseparable.

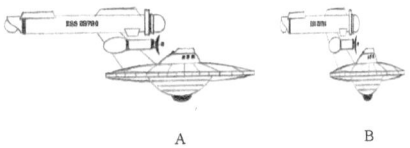

A   B

How a starship would look when stopped. (B) How the same starship might look to observers on Mars as it passes by at 90 percent of the speed of light.

However, our brain functions in the three spatial dimensions—up-down, right-left, and front-back—and time seems to pass linearly. To help understand the problem, try to visualize an object such as a pyramid from the bottom. It looks like a square; but if we move our three-dimensional perspective to the side, it would look like a triangle.

The Modern Concept of Time

One way to try to picture four-dimensional space-time is using the idea of a "time-line." For us it is as though we exist not as a volume in three-dimensional space but as a tube consisting of our volume shifting forward in time from the point of creation (Big Bang) out toward eternity.

This is the modern perception of space and time. It is called "block time." The mixture of space and time, i.e. space-time is absolute and unchanging, much like Newton's three-dimensional space. Things only change in time; they cannot change within integrated space-time. Events do not take place in time; they simply are. The past, present and future are all equally real and the flow of time is an illusion. Since all time is located within any given block or time-line of space-time, the past, present, and future have to all be there and do not change. This modern concept of space-time implies that if we had four-dimensional sight we would see things quite differently. Instead of seeing events unfolding before us with the passage of time, we could witness all time displayed before us in its totality. Theoretically, we could sit in our backyard and witness the dinosaurs that lived there millions of years ago and see the planet Earth being engulfed by our dying Sun. Unfortunately, we do not possess this ability, but if Cosmic Consciousness exists, it transcends space-time, and it could observe all space and time simultaneously. In a way a God-conscious entity would be witness of everything that ever happened, is happening (if there is any meaning to the word), and will happen.

A corollary to the idea of block time is that there is actually no past, present, or future. We only construct these ideas of a flow of time because we view space-time using our three-dimensional mind. However, what happens to the idea of free will if the future is completely determined? How can an individual act in anything less than a predetermined manner if the future is entirely predetermined? The answer is that we make decisions on how to act and do not know how the events will unfold. For us every moment is new. However, for God everything takes place within his mind in the eternity of the now and he knows what decisions we made and will make.

The Meaning of Now

We have seen that time does not tick the same for everyone. When our relative motion is still, all our movement through four-dimensional space-time is through space. However, when we are moving in this medium time slows down. Normally the effect of motion on time is unnoticeable and too small to measure. However, when two atomic clocks are synchronized and one is flown around the globe in a jet the clock in motion is shown to have slowed down slightly just as predicted by theory.

The flow of time can be pictured as the unfolding of snapshots or moments lined up sort of like images on a reel of film. This line of images gives us the time-line. "Now" would be a slice of this line. For observers on Earth that are not moving rapidly relative to one another there is general agreement about now. At the instant of twelve midnight at your home, a clock in India might register twelve noon and this might coincide with a crack of thunder in Australia and a car crash in Japan and an earthquake in Malaysia. Common sense says that we would agree on what lies on a given time slice. It is as though all observers slice the line at the same angle and get a consistent picture of now. However, Einstein showed that motion affects the slice of the time-line differently. When two observers are moving relative to each other there is no longer complete agreement about now. It is as though the motion of the one observer tilts the knife and obtains a different slice of now.

To understand how this works imagine that there are observers on a spaceship several light years from Earth that are moving away from us. This causes their slice of the time-line to be angled toward the past. Their clock is slowed down relative to ours and any information they receive from Earth using regular signals, constrained to the speed of light, will be angled toward our past. Even if the angle is very small, the effect will be magnified by their great distance from Earth. "Now" to them will be events that occurred in our past such as the collision of an asteroid with Earth that already took place. On the other hand, if the distant observers are moving toward Earth, their slice of the time-line, and hence their now will be angled toward our future. They could witness events that have not yet occurred on Earth. All these slices of the time-line are equally valid and real. Each person's perception of now is just as valid as the next no matter whether that person is in motion or not. According to physics all time is out there. All time exists and the now, just like the past and future, are observer dependent and therefore illusionary.

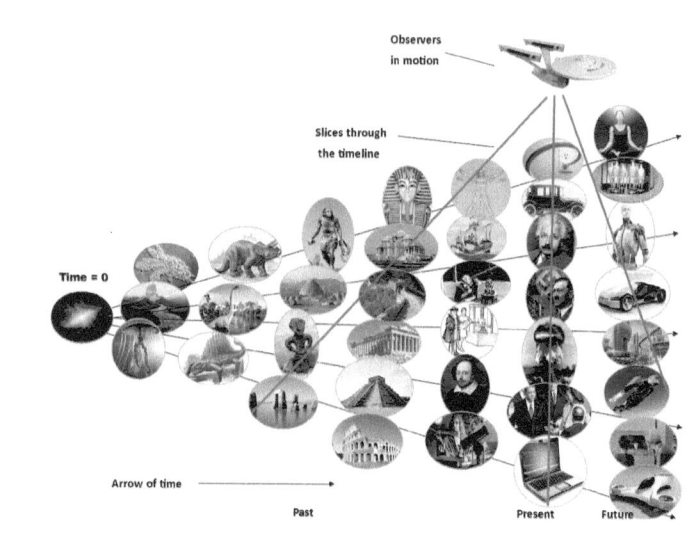

An observer in motion relative to us on Earth gets a different slice of the time-line and consequently experiences a different "now."

Nonlocality of Time

It is safe to say that physics does not understand time. Its equations work equally well without a factor for time or when it is needed whether time flows forward or backward. Earlier we saw how experiments in quantum physics clearly demonstrate that time is nonlocal. That is, an effect can precede a cause.

One of the best examples of this is the delayed choice experiment in which one of the slits creating an interference pattern is closed after the photon has passed but not reached the photographic plate. The interference pattern is lost because the photon somehow knows that it will be closed and it must therefore go to a different location on the photographic plate.

In another experiment, a distant quasar appears to be split into two objects by the bending of light from an intervening galaxy between Earth and the quasar. The light that is bent has roughly fifty thousand light years more distance to travel than the light that comes to Earth directly. However, the photon beams from the quasar interfere with each other in exactly the same way as if they were emitted seconds apart in the laboratory. It appears that the photons remain coupled despite the fact that they were emitted billions of years ago and arrive fifty thousand years apart.

Such experiments have been replicated numerous times and clearly demonstrate: (1) that the wavelike or particle-like behavior of a quantum particle is determined by the decision of the experimenter, (2) an effect can precede a cause or even occur before an apparatus is turned on, (3) that the two complementary aspects of quantum particles, i.e. wave and particle, cannot be simultaneously observed, and (4) that entanglement of quanta occur outside linear time.

The Arrow of Time

We have seen that physics describes time in terms of block time and its corollary time-line. However, if time is really frozen and our perception of past, present, and future is an illusion, why do we universally perceive the passage or flow of time? Does the purely subjective experience that time flows in only one direction have any basis in science? Do we only construct the idea of a flow of time because we view space-time using our three-dimensional mind? Our everyday experience indicates that things are bound by time and seem to follow its rules. Certainly, there must be a basis for time in the laws of physics. The answer to these questions lies in the concept of entropy.

Entropy is a thermodynamic property that is needed to explain certain reactions that occur spontaneously without the expenditure of normal energy such as heat. For example, consider two containers; one filled with oxygen and the other with nitrogen. Now connect the containers by a small tube. Gradually oxygen molecules from container A will pass into container B while nitrogen molecules in B will pass into A. Eventually, the two gases will become completely mixed. The driving force for this mixing is called entropy. The system goes from a highly ordered state (two pure gases) to a less ordered state (mixed gases). Hence, entropy is the driving force causing greater disorder or randomness, and it is ever increasing in the universe.

Our everyday experience of the flow of time is tied inexorably to the constant increase in entropy that we observe. For example, consider a new deck of cards, which comes with all four suits ordered from ace to king plus two jokers. If you throw the deck into the air, it will probably not return to the original order. This is because there are millions of possible ways the cards could fall in a disordered fashion and only one way they could return to their original state. Hence, in all likelihood the cards will become disordered.

One possible exception to the rule of increasing entropy is living systems. They seem to become more ordered with time. For example, consider the difference between an adult homebuilder vs. when he was two years old. As a two year old, he scattered toys and food around the house and now he constructs a house from a pile of building materials. However, living systems do not actually violate the second law of thermodynamics, which states that entropy is always on the increase. When the results of metabolism of highly ordered food molecules are taken into account, the net effect is still an increase in entropy. Ultimately, all the energies utilized by living organisms go back to the energy of the sun, which by converting the mass of hydrogen and other elements into energy, is contributing to an increase of entropy in the universe.

Everything appears to move from order to disorder with the passage of time. This everyday experience creates the illusion of the arrow of time. Entropy is always increasing in the universe. Disorder has increased from the time of the Big Bang forward. This is the explanation of physics for the illusion that time flows. Therefore, if the time-line for the universe is put in reverse we would see an increase in order as we move toward the past, culminating in a point of maximum order at the initial point of the Big Bang. Since, there was no space-time before the Big Bang, the time-line starts there in a state of maximum orderliness. Accordingly, what we call time can only move in one direction from order to disorder.

Time Travel

Is time travel possible? Physics does not rule it out and shows how it may be possible. Motion affects time. We have seen how astronauts might fly off to a nearby star at close to the speed of light and return to Earth many decades later than on their calendar. Gravity also affects time. Gravity distorts space-time. In effect, it can pull on time, slowing its passage. Normally this effect is small, but travel to a black hole with its enormous gravitation pull, and clocks in your spaceship will creep along compared to those on Earth. When you return, you will have aged only a few days while everyone on Earth has aged 50 years and you may be able to greet your grandson who you think should be two years old, but is now actually older than you are.

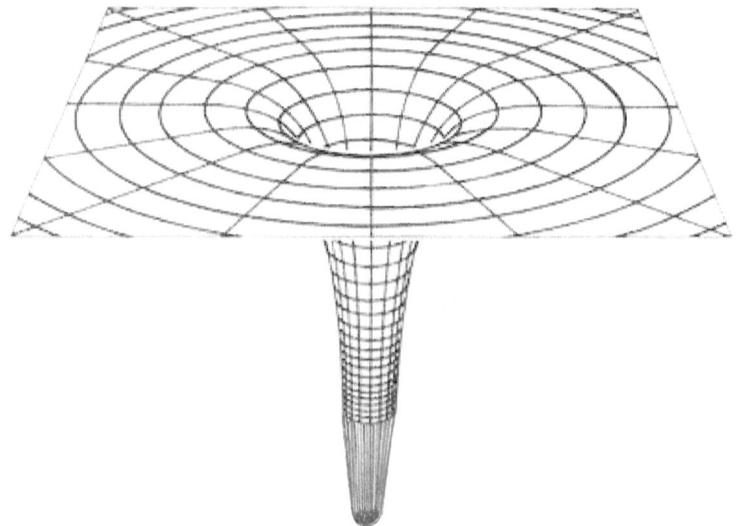

Diagram of the curvature of space surrounding a black hole. The gravitational force of the massive body comprising a black hole distorts the space around it to such an extent that nothing can escape its pull, even light.

Travel to the past might prove to be more difficult. Theoretically, the science of relativity teaches that as an object approaches the speed of light its clock slows down, and at the speed of light the clock stops entirely. Travel faster than the speed of light then the clock would begin to move backward. Hence, if there was a way for humans in the distant future to develop the technology (like the warp drive of Star Trek fame) to travel faster than the speed of light then they could turn back their clock so to speak and return to Earth in the past. However, relativity theory precludes this from occurring since no object possessing mass can even attain, much less exceed the speed light of light.

Another possibility for traveling in time is to find a wormhole that connects your part of space-time with another part. Using a wormhole, you could conceivably travel into the future or to the past. If you find a wormhole that takes you to the past you might arrive back on Earth, meet up with your past self, and tell her all about your future. The laws of physics do not rule out such a scenario. However, it is highly unlikely that any physical object could survive such a journey because physics tells us you need matter with negative density to pass through a wormhole and no such material is known to exist in the universe. Moreover, if time travel to the past were possible, it would create some impossible paradoxes. For example, you could meet your grandfather in the past and kill him. In this case, you would never have been born and therefore could not possibly kill him. In addition, if time travel to the past were possible then we would expect to be constantly bothered by visitors from the future who would no doubt want to come and observe human history first hand.

Consciousness and Time

We have seen that time as defined by science is a relative quality and is intimately entwined with space. If there were no space between objects then it would take no time to go from point A to point B since both points would occupy the same position. Therefore, time can only exist relative to the physical world where objects appear to be separated by space. For distant galaxies, the time for light to reach Earth can be measured in millions of light-years. Anything that distorts physical space, such as gravity or motion approaching the speed of light, distorts time as well. If we could move at the speed of light, we would experience space becoming compressed to a point and time would cease. Such a state would seem to be identical to that described by people who have had a mystical or samadhi experience. This is not surprising since in such a state the bondage of space-time is overcome.

Psychological time is also relative, but it is a separate phenomenon from the space-time of science. In psychological time, there is a clear sense of past, present, and future. Time appears to flow from the past into the future like water passing under a bridge. The flow of psychological time seems to vary depending on our mental state.

Psychological time arises from the movement of our ego consciousness. When we are unconscious or in deep sleep time ceases to exist. Similarly, in the realized or egoless state, our consciousness stops moving and thus time stops. People who have experienced this state of consciousness describe entering the eternal now where both the past and the future collapse into the stillness of the present moment.

Historical events are also relative. An incident that happened on Earth in the last century is a historical event, but as the light or radio waves from Earth will not have reached another planet for a hundred years, on that planet the event has not yet occurred. Thus, historical events depend on time and space and may be interpreted differently by different people. For human beings locked in psychological time the past is gone and has no reality in the now. Similarly, the future does not exist yet and therefore has no reality either. Only the now has reality and in some sense is independent of time.

Author Eckhart Tolle makes the argument in his book *The Power of Now* that to achieve happiness we need to live fully in the now.[2] The past is gone and the future has no existence in the present. When our mind is not fully in the now, he says, we rob ourselves of spontaneity and happiness, and thus all human afflictions can be traced to not living fully in the now. Tolle tells us that we should strive to make psychological time cease by living fully in the now.

TIME ACCORDING TO THE METAPHYSICS OF SPIRITUAL PHILOSOPHY

Modern physics and cosmology demonstrate conclusively that we are part of a far richer and stranger reality than previously conceived. Is there perhaps a way to understand this strangeness by tapping into the ancient wisdom of spirituality? Spirituality is simply the concept that there exists an ultimate reality or Cosmic Entity that transcends the material world and which continuously manifests as reality. The corollary is that mind and matter are derived from spirit (i.e. Cosmic Consciousness). Twentieth-century physicists were not the first to realize that time is an illusion. For millennia seers, saints and ancient texts proclaimed that time is an illusion, for behind this ever-changing reality exists hidden an infinite, unchanging reality. Additionally they claimed that this infinite unchanging reality is at the very core of our being, and the purpose of life is to experience this

Higher Reality and discover not only our connection to God, but also that we are actually God. In doing so we can overcome the three bondages of time, place, and person.

Metaphysics is simply the attempt to understand reality based on the concept of spirituality and the holism that it proclaims. Most physicists try to stay aloof from a metaphysical explanation for the origins and workings of the universe because there is a deep-seated prejudice among scientists against any explanation for reality that cannot be proven by ordinary scientific methods. However, this prejudice should not deter an open-minded person from exploring this alternate explanation of reality. Both the material realism of many scientists and the unity proclaimed by adherents of spirituality begin with an assumption. For materialists there is the assumption that the mass-energy of the universe emerged spontaneously from the void without any cause. On the other hand, the spiritualist assumes that the Cosmic Entity has always existed outside time and has no cause.

There are clear similarities and a few differences between the scientific explanation for creation and that of metaphysics. For example, both ontologies argue that time had a definite beginning. Prior to the formation of space, time did not exist. Time comes into existence at the moment space-time is created. The difference is, that unlike Big Bang theory, which postulates that space-time originated from the nothingness of the void, metaphysics postulates that space-time originated from Cosmic Mind, which in turn originated from the Cosmic Entity. In this way, Cosmic Entity is more subtle than space-time and is therefore outside time (i.e. timeless).

Cosmologists know that space and hence the universe is expanding and that this expansion is not just constant but accelerating. If no new matter were formed then eventually everything in the universe would recede from us at a velocity greater than that of light. In the distant future, people would not see another star in the sky. Finally, after a very, very long time the universe would reach a constant temperature and with no temperature difference to drive any reactions the universe would suffer a "thermal death." At this point of maximum disorder, there would be no movement or change, no events to create the illusion of the flow of time. Hence, for all practical purposes time would cease.

On the other hand, metaphysics says that a thermal death of the universe will not occur. Creation is ongoing and Cosmic Mind is continually undergoing a transformation into space-time, which in turn is continually transformed into additional subatomic and atomic particles (aerial factor).

Since space and matter are continuously being created, new stars will continuously form and therefore there will never be a thermal death of the universe. This continuous formation of new matter is consistent with the hypothesis that there exists a component of matter in the universe that is unseen and unexplained, i.e. dark matter.[3]

Metaphysics explains that the Cosmic Entity witnesses space-time in its entirety. Space-time is pictured as having the same four-dimensional structure as postulated by relativity theory—i.e. block time or time-line, and the Cosmic Entity is a witness of all time since creation. In other words, the creation is an internal psychic concoction of God. As humans we have the capability to witness only the three dimensions of space and our perception of time is limited to a cross section of the time-line that we identify with the "now." In contrast, the Cosmic Entity has the capability to see the fourth dimension, which would be analogous to witnessing the line in totality—i.e. not just a single line, but all time-lines (i.e. the cosmic-line) emanating from the Cosmic Nucleus or starting point of creation—hence, an omni-view of all space and time.

For God everything is known since it all takes place within his mind in the eternity of the now. When we as humans surrender the idea that we are separate from God (enter the egoless state) then we realize that it is actually he that performs all actions. Once the illusion of separateness from God is dispelled then the illusion that time flows is also lost since there is the realization that change is an illusion. Within the wholeness of God, no change actually takes place.

Metaphysics also solves many of the anomalies and enigmas that physics currently cannot explain. One of these is the problem of how, against all odds, a flat universe emerged from the chaos of the Big Bang. A flat or long-lived universe is required for the evolution of advanced beings but is highly unlikely to occur by chance.

Secondly, physics has no explanation for why more matter particles emerged from the Big Bang than antimatter particles. All the theories indicate that in a universe born in a tremendous explosion such as the Big Bang, there should be equal amounts of matter and antimatter. If this were the case then the colliding pairs of matter and antimatter particles would annihilate each other leaving nothing but empty space. For some unexplained reason we have a universe that initially had an excess of matter particles over antimatter particles.

There is also fine-tuning of certain cosmic ratios and universal constants that are required if conscious life is to evolve in our universe. For example, the mass of elementary particles, their relative abundances, and the forces that exist between them are mysteriously adjusted to favor a universe with long-lived stars. Key physical parameters and constants appear to be exquisitely fine-tuned to allow for the existence of stars, planets, water, and ultimately for the emergence of living organisms and self-conscious beings. The four fundamental forces responsible for shaping our universe—gravity, electromagnetic, weak, and strong nuclear forces of nature—seem to be finely tuned in exactly the right way to enable the universe to exist precisely as we observe it. If any one of the four forces differed by even a few parts per thousand it would have had a profound effect upon the nature of our universe, making life as we know it impossible.

Metaphysics has a simple explanation for all of these problems—the universe is the product of conscious design. The universe did not originate by random, chaotic processes, but by the careful design of the Cosmic Entity.

Physics also has no suitable explanation for the nonlocality of space-time, or how the mind of the observer affects physical reality. Metaphysics states that the universe is a Singularity or indivisible Whole, within which everything is connected. Nonlocality is a result of this unity. Consciousness can and does affect matter because matter is an epiphenomenon of mind and consciousness. If mind and consciousness were epiphenomena of the brain (matter), there would be no mechanism by which mind could have an effect on matter.

The nonlocality of time refers to the idea that before/after and cause/effect is meaningless since all time is present within the framework of four-dimensional space-time, and is unchanging. Both metaphysics and physics agree that in a sense the future already exists. For science, this is required from the integration of space and time and the speed limit for light. For metaphysics, it is required by the fact that for physical matter and energy there must be a limit on how fast they can move. If there was no speed limit for the physical components of creation then in theory they could reach infinite speed which would be analogous to being everywhere at the same time. Only the Cosmic Entity can have this capability. The speed limit for light is an insurmountable wall separating mind and matter and prevents matter from traveling to the past, which would create impossible paradoxes, and prevents matter from being converted back to mind. Hence the hierarchy: consciousness » mind » matter is maintained.

In order to be a viable explanation for reality, metaphysics (i.e. holism) should be able to explain all the many bizarre aspects of time and space, in a logical and rational way, but also how and why conscious life evolved in the universe.

In Chapter 4, we will explore the Cycle of Creation and see how most if not all these questions can be answered by assuming that creation begins with consciousness and that the material world is actually created from consciousness.

3

Simple is Better

OCKHAM'S RAZOR

The fourteenth-century English theologian William of Ockham is credited with first postulating the principle known as Ockham's razor. The principle states that when there is more than one competing hypothesis to explain some phenomena, both of which are equally plausible, it is the simple explanation that is normally correct. Ockham's razor may be expressed in Latin as: *pluralitas non est ponenda sine necessitate* (plurality should not be posited without necessity). In other words, the principle recommends selection of a hypothesis that introduces the fewest assumptions and entities while still answering the question over one that uses multiple assumptions, entities, and complexities.

This razor is best used as a guide to help in the development of theories or hypotheses rather than as a scientific law. The main postulate of Ockham's razor that simpler is better is a reflection of the Unity Principle that the diversity we witness in the universe masks an underlying oneness. Therefore, the complex phenomena that occur in our universe will ultimately be explained by a simple unified theory of everything that expresses the connection between matter, energy, mind, and ultimately Consciousness.

In the paragraphs that follow, we will explore complex versus simple explanations for several observed phenomena. In each case, the simple explanation is in keeping with the postulates of the Unity Principle while the complex explanation has often been contrived in order to avoid a metaphysical explanation.

Consciousness is a Product of Matter

The doctrine of material realism assumes that consciousness, which is subtle by nature, is a product of matter, which is crude. In other words, mind and our awareness of it are a product of unconscious, non-living matter-energy. Consciousness depends on mind, which depends on a nervous system, which depends on the evolution of more developed life forms.

This extremely complex process of evolutionary development would depend first on the chance occurrence that molecules capable of reproduction developed spontaneously to form living organisms. Secondly, that these organisms were capable of developing complex structures via point mutations of their DNA by a selection process that rewarded organisms that were better able to adapt to their environment. Finally, that the brain, which is a crude object, was capable of developing mind, simple awareness, and consciousness of itself.

While it is true that the mind expresses itself in different parts of the brain through different kinds of sensory functions and thought-waves, scientists have never been able to locate any physical structure in the brain that is responsible for conscious self-awareness. If consciousness were actually a product of the brain then it would make sense that there must be some physical structure responsible for it; in other words, that there exists a brain basis for the sense of self. However, no such physical brain structure has been found. Hence, a simpler and more logical explanation is that mind and matter are a product of Consciousness. In other words, the crude has evolved from the subtle, and life and resulting unit consciousness evolve naturally on a planet when conditions allow it.

Nonlocality and Quantum Connectivity

Experiments have demonstrated conclusively that quantum fields are nonlocal in both space and time. For example, paired photons are connected or entangled no matter how far apart they are in space—even if they are on opposite sides of the universe. These phenomena of space-time nonlocalities point to an undivided wholeness in the cosmos and strongly suggest that there is an information field that is subtler than space-time. This subtle field connecting material objects can be termed mind or consciousness.

The Unity Principle provides a very simple explanation for these phenomena. A more complex explanation is the many-worlds interpretation of quantum mechanics. According to this theory, all the superpositions (quantum possibilities) are real. When there is a measurement or observation, we create another possible direction that one of the possibilities can take and a new universe is created. Hence, every time an observation is made the world branches, and there is another copy made, and therefore each of us exists in countless worlds, all real, but slightly different.

In this many-worlds ontology, the potential quantum outcomes exist throughout space-time and are therefore connected. When an observation of one photon of an entangled pair is observed a new universe is created in which the other paired photon has opposite polarization. The many-worlds interpretation may be used to explain quantum nonlocality, however, it relies on the fabrication of an almost infinite number of parallel universes and is untestable since theoretically there is no interaction between worlds.

The many-worlds explanation for quantum nonlocality embodies complexity at its most extreme and is the antithesis of Ockham's razor. Physicist Paul Davies objects to this theory not only because of its unnecessary complexity, but also because it is hard to see how law and rationality can emerge from total randomness.[1] It appears that physicists concocted the theory in an attempt to avoid the simple explanation that there exists a metaphysical oneness to the universe.

PSYCHIC ABILITIES

Skeptics dismiss the entire body of evidence indicating that humans possess psychic abilities as inconclusive, faked, or a product of flawed science. Yet, there have been thousands of well-controlled scientific experiments that have clearly demonstrated psychic abilities with high statistical significance. Experiments such as those affecting computerized random number generators (telekinesis), reading hidden cards (clairvoyance), and sending mental images (Ganzfeld telepathy experiments) are a few examples. When statistical meta-analyses are done on similar studies, the results demonstrate conclusively that there is a real effect, although it is small in most individuals.[2]

To dismiss this huge body of evidence, one must either deny the statistics; assume that there has been a major effort to defraud the scientific community; assume that virtually all the studies were improperly controlled; or assume that for every published study showing positive results there were hundreds that showed no effect and were not published. This is a complex explanation for the data, concocted by skeptics who are convinced that mind is localized and wholly a product of the brain. A far simpler explanation for the enormous body of evidence demonstrating the psychic abilities of human beings is that these abilities are real and are a product of the nonlocal unconscious mind that allows humans to perceive and affect physical objects beyond their own body.

Occasionally a person comes along who can be considered a psychic savant. One such individual was Edgar Cayce. Born in Kentucky in 1877, Cayce received only an eighth-grade education before he had to work on the family farm. At the age of twenty-three, after suffering from chronic laryngitis, Cayce sought help for his condition from a hypnotist. He was told to perform self-hypnosis in order to obtain a permanent cure.

After entering a self-induced trance, he described in detail the ailment and its cure. This was his first psychic reading and over fourteen thousand were to follow before his death in 1945. He performed his readings while in a trance and after awakening had no recollection of what he said. The majority of his readings were for people seeking medical diagnoses and treatments. All he required was the individual's name and location. He would give an accurate description of the illness or physical problem as though he had X-ray vision and then offer a treatment for the condition. The accuracy of his readings was astounding while the terminology he used was not that of an unschooled individual but of someone highly trained in medicine.

Cayce also did readings on geology, chemistry, physics, electricity, history, political science, sociology, and anthropology. In his metaphysical readings, he confirmed the validity of the Law of Karma and reincarnation, astrology, psychic abilities, and the power of meditation. He also made numerous prophecies such as the stock market crash of 1929 and the start and combatants of World War II that turned out to be uncannily accurate. The simple explanation for Edgar Cayce's psychic abilities is that while in a trance he tapped into Cosmic Mind. There has never been any other reasonable explanation for Edgar Cayce's incredible psychic abilities.

Homing and Other Animal Instincts and Behavior

Many species of animals exhibit homing behavior that seems to defy rational explanation. Examples include salmon, many species of birds, and some insects such as monarch butterflies. Scientists have posited complex explanations for such abilities such as magnetic navigation, sun navigation, and sense of smell or a combination of such sensory cues. However, the magnetic field of Earth changes constantly over time and its position changes over the surface of the planet. It is also affected by sunspots, iron deposits, and magnetic anomalies and is so "noisy" that it is unlikely that it could be used to provide anything but a general direction to a homing animal. Sun navigation depends on both surface position and time. Animals would have to have an extremely accurate internal clock to utilize this form of navigation. Mariners can obtain their latitude using a sextant and an accurate watch, but longitude is not easily measured using the sun. Therefore, navigation based on the sun can at best provide homing or migrating animals with a general direction—not precise location.

A keen sense of smell has also been offered as a possible explanation for homing instinct in animals such as salmon and dogs. Yet, it is unlikely that a particular stream would maintain the same smell signature after many years had passed between the time the salmon were hatched and the time when they returned to their spawning beds. Similarly, there have been many reported cases where a dog found its way to its owners at a new location weeks and even months after it was lost. This would suggest that if the dog were using its sense of smell to find its way home then discrete smell molecules linger in the environment much longer than can reasonably be expected.

The homing ability of some animals is a confounding mystery to scientists who postulate a sensory explanation for such behavior, including the possible utilization of several senses simultaneously. The simple explanation for such behavior and ability is that the animals are guided by instinct and that animal instinct is a product of Cosmic Mind.

It is easy to explain how the physical attributes of animals are passed on between generations, but it is not obvious how complex instinctual behavior in animals is passed on. DNA is the genetic material responsible for the transmission of the physical structures of all living organisms (with the exception of some viruses that utilize RNA). It is the blueprint for constructing the entire organism starting from a single cell. The principal role for DNA is to code for proteins. It is difficult to understand how the structure of various proteins can account for complex instinctual behavior.

By some accounts only about 3 percent of our DNA is utilized for protein coding. What the other 97 percent is for is unknown and it is sometimes called "junk DNA." Some scientists have speculated that it carries inherited and instinctual memories from one generation to the next. However, there is no evidence directly linking DNA with memory. If there were "genetic memory" then it would be reasonable to expect that we would inherit knowledge from our parents, grandparents, etc. Yet such knowledge is lacking and there is no known repository of inherited knowledge in either the genome of humans or any other animal.

Many animals show extremely complex inherited behavior. For example, the courtship display of the Bird-of-Paradise (*Lophorina superba*) is incredibly bizarre and elaborate. Like other instinctual behavior in animals, such courtship displays are inherited—not learned. A species of spider spins a web of a specific and complex design that it has not learned from a parent. How and why do all such spiders produce a web of exactly the same design? One explanation is that such unlearned behavior is passed from one generation to the next by a complex yet unknown mechanism involving DNA, proteins, or RNA. However, a good portion of animal behavior is extraordinarily complex and cannot be fully explained based on simple genetics. The simple explanation is that animal instinct is governed by the Cosmic Mind.

Throughout history, there have been reports of animals' "sixth sense" in detecting hurricanes, earthquakes, tsunamis, and volcanic eruptions before the event happens. Examples include dogs and cats and other animals acting strangely, birds taking flight and rats leaving buildings before the ground begins to shake. This mysterious ability may allow some animals to "sense" geophysical changes in the earth before they happen.

For example, it was reported that before the giant tsunami hit the coasts of Sri Lanka and India on December 26, 2004, elephants screamed and ran for higher ground, dogs refused to go outdoors, zoo animals rushed into their shelters and could not be enticed to come back out, and flamingos flew to higher ground.[3] At the hard-hit Yala National Park in Sri Lanka, stunned wildlife officials reported that hundreds of elephants, leopards, tigers, wild boar, deer, water buffalo, monkeys, and smaller mammals and reptiles had escaped unscathed.[4] Rescue workers found surprisingly few dead animals at the park.

There was at least one example where authorities successfully forecast a major earthquake, based primarily on the observation of the strange behavior of animals. In 1975, Chinese officials ordered the evacuation of Haicheng, a city with one million people, just days before a 7.3-magnitude quake. Only a small portion of the population was hurt or killed because of the evacuation. It was estimated that the number of fatalities and injuries could have exceeded 150,000 if no warning had been given.[5]

Some animals seem to be able to sense that a person is about to die or have a medical condition or emergency. A recent example of this was reported in the July 2007 issue of the *New England Journal of Medicine* about a cat named Oscar that seemed to be able to "predict" the deaths of patients in a nursing home in Providence, Rhode Island. Just before patients died, Oscar would sit down by their beds and would become very upset if forced out of the room before the patient died. The article sited at least twenty-five successful predictions of impending death by the cat. Dogs have been trained to give warning of an impending epileptic seizure in people prone to this condition. They are specifically trained to give warning to their owners so that they can take appropriate precautions before the seizure strikes. Whether the dog smells something from the person that provides a clue to an impending seizure or just uses its "sixth sense" is unknown. Dogs are also known to be able to detect certain cancers in people even at an early stage. It has been postulated that it is their keen sense of smell that allows them to detect malignancies in humans, but this is only speculation. The simple explanation for the uncanny ability of animals to sense things beyond the purview of the five standard sense organs is that they are intimately connected to the subtle and pervasive Cosmic Mind.

THE GOLDILOCKS ENIGMA

Scientists today have begun to recognize that the fundamental forces and physical laws and constants all seem to be exquisitely fine-tuned to allow for the existence of stars, planets, water, and ultimately for the emergence of living organisms and self-conscious beings. As noted in Chapter 2 there are three fundamental forces responsible for shaping our universe: gravity, electroweak, and strong nuclear force. The electroweak force can be divided into electromagnetic and weak forces for the purpose of this discussion.

What is most intriguing is that these four forces of nature seem to be finely tuned in exactly the right way to enable the universe to exist precisely as is does. If any one of the four forces differed by even a few parts per thousand it would have had a profound effect upon the nature of our universe, making life as we know it impossible. This mystery of why the forces of nature seem to be perfectly tuned to allow for the development of conscious life has been termed the Goldilocks enigma.

The exquisite fine-tuning of forces is seen in many facets of nature. For example, the strong nuclear force, which binds protons and neutrons in the nucleus of atoms, is just the right strength to make stable atomic nuclei and allow the fusion of hydrogen nuclei in the center of stars. This releases just the right amount of energy to power stars for a very long time, thus maintaining a more moderate and constant temperature on the planet and enabling life forms to evolve. Similarly, the weak force is perfectly tuned to allow for nuclear decay in heavy elements, which heats Earth's molten core resulting in volcanism and a magnetic field that shields life on Earth from deadly cosmic radiation.

The electromagnetic force is responsible for light, heat, electricity, and for the construction of matter, as we know it. Even a slight variation in its strength would have a profound effect upon the chemistry of life.

If the force of gravity differed even slightly from its actual value, models predict that stars, galaxies, planets, and hence living organisms could not have developed from the dust left by the primordial Big Bang explosion. Even the formation of matter following the Big Bang depended on a tiny excess of matter particles over antimatter particles. Luckily, for us there was just the right excess of matter particles created during the inflation phase of creation.

Stephen Hawking has this to say about this enigma:

> Why is the universe so close to the dividing line between collapsing again and expanding indefinitely? In order to be as close as we are now, the rate of expansion early on had to be chosen fantastically accurately. If the rate of expansion one second after the Big Bang had been less by one part in ten billion, the universe would have collapsed after a few million years. If it had been greater by one part in ten billion, the universe would have been essentially empty after a few million years. In neither case would it have lasted long enough for life to develop. Thus one either has to appeal to the anthropic principle or find some physical explanation of why the universe is the way it is.[6]

Physicist Roger Penrose has noted that the "fine-tuning" necessary for life to emerge from the Big Bang in terms of phase-space-volume is so absurdly unlikely to have occurred by chance that there must be another explanation.[7] Sir Martin Rees of the Royal Observatory in Britain writes that there are just six numbers necessary for our emergence from the Big Bang as conscious beings. However, if there were a minuscule change in any one of these cosmological constants then life and the universe as we know them would be impossible.[8]

Many other scientists have pointed out how incredibly fortunate we are to have a universe that allows for the development of living organisms. They have even coined a term for it: the anthropic principle. This principle states that the universe must be exactly as it is in order to allow for the emergence of intelligent life to observe it. At the heart of this idea is the allowance for multiple universes (or multiple Big Bangs)—multiverse theory. According to multiverse theory, during the chaotic inflation phase immediately following the Big Bang an untold number of universes were created, each with slightly different rates of expansion, cosmological constants, and forces. Out of the almost infinite universes created, ours is the one with just the right conditions for the development of conscious beings.

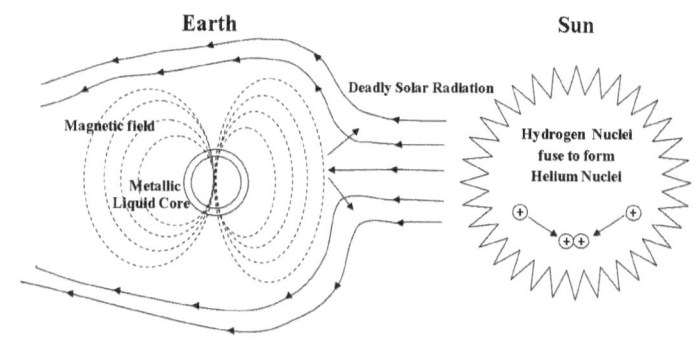

The Goldilocks enigma. All the forces of nature must be "just right," precisely balanced to allow for the emergence of life on Earth

As the many-worlds hypothesis used to explain quantum nonlocality, multiverse theory is untestable since the other universes are unobservable. The theory draws upon extreme complexity in order to explain the Goldilocks enigma. The simple explanation that many scientists scoff at (probably because it sounds like metaphysics) is that the universe began as a singularity and has remained as such from the time of the Big Bang onward. The universe is exactly tuned for the development of conscious beings because it is created from Consciousness. Hence, a better definition of the anthropic principle might be: we are conscious of this universe because it was created from Consciousness in such a way as to allow for the development of individual conscious beings by a process that can be called "conscious design."

A Theory of Everything

A grand unifying theory of everything would go a long way toward simplifying physics. Quantum theory does a good job of explaining the interactions of matter and energy on a tiny scale, while Einstein's theories of relativity work well on the large scale. However, the two theories have not been successfully tied together into a unified theory. Similarly, the Standard Model or Grand Unified Theory (GUT) of physics has been successful in tying together the strong force, weak force, and electromagnetism, and the particles that govern these forces. The only missing element in this model was the Higgs field, which was recently shown to exist by experiments utilizing the Large Hadron Collider.

However, to date, all attempts to draw gravity into this model have failed and the great minds in physics have all but given up on trying to bring gravity and its unseen graviton into the fold. If Ockham's razor is applicable and the universe does exist as a Singularity then it is logical to expect that scientists will eventually discover a mathematical model that explains how all the forces and particles found in nature are interconnected.

At least one physicist believes he has discovered a mathematical model that would explain everything. His name is Garrett Lisi and he calls his model an "Exceptionally Simple Theory of Everything."[9] It is a unified field theory combining the Grand Unified Theory of particle physics with the relativistic description of gravity. Lisi visualizes a model consisting of a Lie-group, a mathematical shape consisting of a collection of circles twisted around each other in a specific pattern. His model utilizes an E8 Lie group consisting of 248 circles.

Each circle can be associated with a different kind of elementary or force particle. He also found a set of circles that seemed to act like the elusive graviton. For the first time a physicist has come up with a model of how gravity might fit in with the other forces. This model could explain how all the standard particles are unified as well as all the fundamental interactions including gravity. The theory predicts several new particles and will only succeed if some or all of those particles are discovered. Nonetheless, the incredible simplicity and symmetry of his model suggests that physicists may someday develop and obtain evidence for a simple theory of everything.

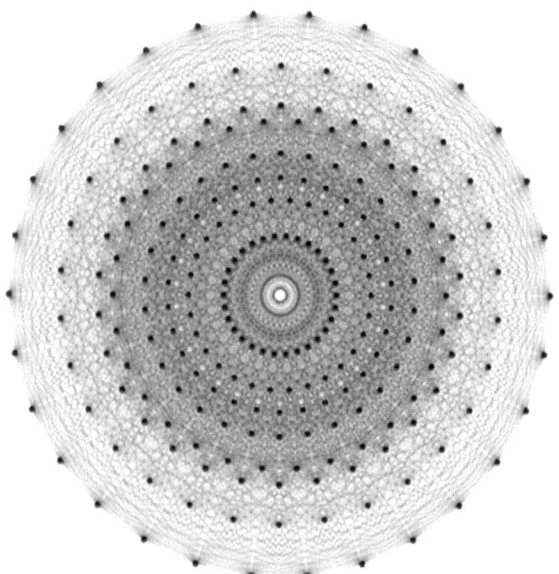

An E8 Lie group. A perfectly symmetrical 248 dimensional object. Courtesy of the Atlas of Lie Group Project.

EVIDENCE FOR CONSCIOUS DESIGN

Living organisms show incredible diversity and complexity. Many of the complex structures observed in living organisms seem to point to an evolutionary process that utilized design, i.e. teleological evolution. For example, the blood-clotting process in vertebrates is exceedingly complex and the whole system needs to be in place before there would be an advantage conferred upon offspring. The term coined by biochemist Michael Behe for such a system is "irreducibly complex."[10] If a biological structure or system is reverse engineered and requires all of its parts to function then it can be termed irreducibly complex.

Besides the blood clotting system in vertebrates, many other biological structures may be considered irreducibly complex. Biologists have difficulty explaining how complex systems such as the eye, bacterial flagellum, and the immune system evolved by random mutations of DNA followed by natural selection. Natural selection with its "survival of the fittest" requires that the system or structure be somewhat functional and therefore advantageous to an organism before it would be conferred upon and propagated by offspring of that organism.

Hence, if a system has many interconnected parts in order to function, it is difficult to explain how it evolved in a stepwise manner. The complicated answer for how complex organisms developed is that evolution takes place by chance mutations and natural selection. The simpler explanation is that evolution occurs as organisms struggle to survive, while underneath this struggle is the guidance of the Cosmic Mind.

When we look at the incredible beauty, complexity, and exquisite fine-tuning of our world it is natural to think that it is the product of design rather than the accidental happenings of blind chance. Throughout recorded history and in almost every culture people have looked in wonder at the world and believed in the idea of a Grand Designer or God. Material realists have offered very complex and unproven theories for how our universe and conscious beings arose from the dust of the Big Bang. If Ockham's razor were applied to the materialists' ideology then it would have to be rejected in favor of the simple explanation of the Unity Principle—that everything in the universe is created from Consciousness.

The Universal Mind of Man

Human beings may also be connected to Cosmic Mind, although the connection, if it exists, must be obscured by the subconscious and conscious minds. In Chapter 7, we will discuss the layers of the human mind.

According to this scheme, the unconscious or superconscious mind is a collective mind shared by all unit minds.

Carl Jung called this level of mind the "collective unconscious." His evidence for the existence of this universal mind linking all humankind was based on the observation that among cultures with no historical influence upon one another the basic symbols and myths were universal. Jung called the predisposed patterns for particular myths and symbols "archetypes." He argued that they exist in the collective unconscious of humankind and are similar to animal instincts in that they influence the basic structure and organization of the human psyche. Jung identified several archetypes, such as the hero, the mother, and the devil.

He also offered numerous examples of myths and symbols that arose in cultures with no direct link to one another. One example of a pervasive symbol is the cross and another is the twisted cross or swastika. The word "swastika" is derived from Sanskrit: *su* (good) + *asti* (being) + *ka* (object). Hence, it symbolizes well-being, good luck, or spiritual success. Besides being an ancient symbol of Hinduism, Buddhism, and Jainism, it has been found in Neolithic Europe, Asia, Africa, the Middle East, and China. It was also used by Celts, Greeks, Roman, Germanic and Slavic tribes, and by Native Americans. Archaeological evidence suggests that it served as a good luck charm or religious symbol in these various cultures.

Joseph Campbell gives many other examples of universal symbols and myths in *The Mythic Image* and *The Hero with a Thousand Faces*. The existence of Universal Mind provides a simple explanation for why these archetypal patterns exist. Materialistic skeptics have offered no other meaningful explanation for these phenomena.

4

Brahma Chakra: The Cycle of Creation

MONISTIC RELIGIONS AND SPIRITUAL TRADITIONS

The metaphysics of unity is found in many ancient religions and spiritual traditions. This view of reality was not based on direct observation of quantum particles but on mystical experience. Mystical experience is a non-intellectual or intuitive experience of reality. Although these experiences appear to differ in the details, they share one underlying and fundamental element—the view or experience of a holism or oneness of the cosmos. The experience that the universe is One Organic Whole is a common thread found in the writings of the mystics even though these individuals differed greatly in their historic, geographic, and cultural traditions.

The great religions that grew from the mystical visions of such men as Sadashiva, Lao Tzu, Buddha, and Baha'u'llah, include Vedanta/Hinduism, Taoism, Buddhism, and Baha'i. These are the primary monistic religions of the East. Other more largely spiritual traditions of monism include Tantra, yoga, Sufism, and Hasidic and Kabbalistic Judaism. The goal of these religions and traditions is for the individual to merge or become one with the Whole or God. This becoming one with the Cosmic Entity is sometimes called enlightenment, self-realization, or liberation—implying freedom from ties of individuality or ego. The state or experience of oneness with the Cosmic Entity or God is sometimes called Cosmic Consciousness, unqualified liberation or salvation, samadhi, nirvana, moksha, mukti, or satori. The One has many names or terms, including

Tao, Brahma, *Paramapurusha*, Paramatman, *Bhagavan*, Allah, Jehovah, and God.

Brahma the Cosmic Entity

The Vedas or ancient religious scriptures of Hinduism call the Cosmic Entity Brahma. Brahma is synonymous with the western concept of God. However, in western religions God is often portrayed as separate from his creation while the eastern concept of Brahma has him as both the creator and the created. According to Tantric and Vedic philosophy, Brahma has two complementary aspects: *Purusha*, or Consciousness, and *Prakriti*, the Qualifying Principle. Like other complementary constructs, such as wave-particle, *Purusha* and *Prakriti* can be considered as two sides of a coin—the coin representing Brahma.

Without the action of *Prakriti*, the unqualified *Purusha*, or Cosmic Consciousness, could not undergo transformation and would remain unmanifest as pure awareness.

The Qualifying Principle, *Prakriti*, has three basic modes with which it binds or qualifies. The subtlest mode of *Prakriti* is *sattvaguna*. Here *sattva* means "sentient" or "weak" and *guna* means "binding force." The second mode of *Prakriti* is *rajoguna*. *Rajo* means "mutative" or "active principle." The strongest binding force of *Prakriti* is *tamoguna*, the static principle. Normally the Cosmic Entity, Brahma, remains in an unqualified state: *Nirguna* Brahma (literally, without *guna*). The binding forces of *Prakriti* can be thought of as being disorganized or dormant and thus unable to act upon *Purusha*.

In Tantra, creation is thought to begin when the three binding forces of *Prakriti* form a perfect equilateral triangle and capture a portion of the unqualified *Purusha* or Cosmic Consciousness in the interior of the triangle. The *Purusha* cannot withstand being confined in this manner and bursts forth from one of the vertices of the triangle beginning the process of creation. This explosion of Consciousness from what is to become the Nucleus of Creation is analogous to the beginning of the cosmos envisioned by astrophysicists—the so-called Big Bang. The movement of Consciousness, which begins in this Cosmic Nucleus, is extroversive or centrifugal in character and undergoes a change from subtle to crude. *Saincara* is the Sanskrit name given to this particular movement in the Cycle of Creation.

The Qualification of Brahma

The Consciousness or *Purusha* that bursts from the Nuclear Point of creation is not qualified but it does come within the scope of *Prakriti*. This aspect of the Cosmic Entity is said to be *Saguna* Brahma (literally, with *guna*). The unqualified *Purusha* can be thought of as analogous to a wave with an infinite wavelength, hence a straight line. This unqualified *Purusha* is pure being or awareness without even the idea of its own existence.

When a portion of the unqualified *Purusha* comes under the influence of *Prakriti's sattvaguna*, however, the feeling of "I exist" is created. This is analogous to creating a slight curvature in the straight-line wave of the unqualified *Purusha*. This sense of "I exist" or "I am" is called *Mahattattva* in Sanskrit and is the Cosmic "I." This is the first qualification or bondage of the Cosmic Consciousness or *Purusha*, but it is a very subtle bondage that is imparted to only a microscopic portion of the unqualified Consciousness.

When a portion of the *Mahattattva* comes under the influence of the *rajoguna* force of *Prakriti*, the *Ahamtattva* or "I do" feeling is created. Naturally, the wavelength of the *Ahamtattva* is less than *Mahattattva* since it is more qualified or bound. This Cosmic "doer I" is subjective in character and has no objectivity; like the *Mahattattva* it is more or less a theoretical construct.

The Dominance of Static *Prakriti*

As the *tamoguna* or static principle of *Prakriti* begins to dominate the Creation Cycle it first acts upon a portion of the *Ahamtattva* and creates the objective Cosmic "I" or *Citta*. This Cosmic *Citta* has objective reality but no material reality since matter is not yet formed in the *saincara* phase of creation. The objective Cosmic Mind-stuff or *Citta* is somewhat analogous to the objectification of an idea and can be compared to a scene in our minds. For example, suppose we close our eyes and try to visualize a person riding a horse. There must be willfulness or doership associated with our sense of "I" as our mind takes on the colors and forms of the horse and rider. For us the scene has only subjective reality, much like a dream or hallucination. It has no objective reality since we lack the power to create a living horse and rider.

This mental picture is analogous to the *Citta* or objective Mind-stuff, except that the *Citta* is Cosmic in proportion. These three aspects of creation, *Mahattattva, Ahamtattva,* and *Citta,* together constitute Cosmic Mind and they represent the first phase of Brahma Chakra (literally, the circle of God).

The process of *saincara* continues under the gradually increasing dominance of *tamoguna*. As a result, a portion of the Cosmic Mind-stuff or *Citta* is transformed into *akasha* or ethereal factor, which is essentially space-time. Recall that early scientists observed wave-like properties of light such as diffraction and thought that light waves reaching our planet from distant stars needed a medium through which to travel (just like all waves). They labeled this medium "ether." The existence of a subtle medium for the transmission of light waves was disproved when it was discovered that light traveled at the same speed whether it came from a luminous object that was traveling toward Earth or away from Earth. In addition, it was discovered that light behaved as a particle (photon) as well as a wave. A particle of light would not need a medium for travel in space. Although the etheric medium theory was disproved, for the purpose of this discussion the term "ether" or *akasha* is used synonymously with space-time.

Most non-scientists assume that space consists of a vacuum, which has no physical matter and no energy. However, quantum mechanics and relativity theory require that a vacuum must contain an almost unlimited quantity of energy—much more than matter as defined by Einstein's famous equation: $E = mc^2$; where energy (E) is equal to mass (m) times the speed of light in a vacuum (c) squared. Hence, the ether/space has tremendous energy locked within it, and it is very important for understanding the true nature of the cosmos.[1]

Mystics proclaim that the subtle vibration of ether (*akasha*) can be heard in the mind as a sound. They call this sound Om and proclaim it the emergent sound of creation.

The transformations of *Purusha* or Consciousness into "I feeling" (*Mahattattva*), "doer I" (*Ahamtattva*), "subjective I" (*Citta*), and ether (*akasha*), take place outside the scope of time or within the first moments following the explosion of Consciousness from the Cosmic Nucleus or singularity of the Big Bang. In these first few seconds, space (*akasha*) expands rapidly from the Nucleus resulting in a rapidly expanding physical universe.

As the pressure of the static binding principle, *tamoguna*, continues to increase, it transforms a portion of the *akasha* into the aerial factor or the simplest form of matter. Enormous amounts of subatomic particles were created in the first few seconds of creation and became spread out during the initial expansion of the universe. Initially these particles composing the aerial factor were too hot to unite and form atoms. This hot plasma also emitted radiation or luminous factor.

Evidence obtained from radio telescopes and the Planck spacecraft indicate that even today we can observe a relic of the Big Bang termed "cosmic microwave background radiation." Studies of this radiation show that it is not completely homogeneous, which may have resulted in the inhomogeneity of the universe. Thus, today most observable matter is concentrated in distant galaxies. As the universe expanded and cooled, these subatomic particles, such as protons, electrons, and neutrons, coalesced and formed principally hydrogen gas. Hydrogen is the simplest element, having one proton and one electron. It is possible that additional hydrogen is still being formed from the ethereal factor long after the Big Bang first created space-time.

Under the relentless pressure of static *Prakriti* (*tamoguna*), the hydrogen atoms, which were not uniformly distributed in space, become attracted to one another by the pull of gravity. Once a huge mass of hydrogen forms, the gravitational force is so strong that it causes the hydrogen atoms to become incredibly compressed and lose their electrons. Under these conditions of extreme heat and pressure, the nuclei or protons begin to fuse, igniting a baby star. This process is termed thermonuclear fusion and is the same process that occurs in all stars, including our Sun. It is also responsible for the enormous energy released by a hydrogen bomb.

The energy is released when the hydrogen nuclei fuse to form a helium nucleus of two protons, and in the process, some mass is converted to energy. Even though the loss of mass is small, the energy is large since in Einstein's equation relating mass to energy—$E=mc^2$—"c" (the speed of light in a vacuum) is a very large number. This means that even the small mass of a proton has a large amount of energy locked within it. This process creates the luminous factor (*tejastattva*), which is known scientifically as electromagnetic radiation.

A typical star will consist principally of hot plasma, a form of luminous gas in which most of the atoms are ionized or do not have electrons directly associated with atomic nuclei. Besides helium, the second element, thermonuclear fusion creates small amounts of heavier elements, but none heavier than iron. For heavy elements to form, such as copper, zinc, silver, gold, etc., a massive star has to undergo a stellar explosion known as a supernova.

Such events are rare but can emit as much energy in a few weeks as our Sun will put out during its lifetime. In addition, supernovas throw off enormous amounts of gas and create the heavier elements, some of which are necessary for life. Hence, life as we know it would be impossible if we did not have within our bodies the dust from exploded stars. The shock wave from a supernova can also trigger the formation of new stars. Stellar material is constantly being recycled and it is not unusual to look into the night sky and observe new stars forming from the dust of dead stars.[2]

As a star forms from gaseous hydrogen it will often develop what is termed a gaseous protoplanetary disk. After a substantial amount of time hot gases in this disc cool with the accretion of dust and heavier material produced by the star or attracted to it by its massive gravitational field. Eventually these gases coalesce to form protoplanets. At first, the protoplanet will exist in a molten state, but over time, it will cool and develop a solid crust. Thus, the liquid and solid factors arise by the continued binding force of *tamoguna* on the gaseous and luminous plasma that makes up a star. The solid factor is the crudest form that can be made from Cosmic *Citta*. In other words, in a solid the pressure of the static *Prakriti* has reached its maximum.

In the solid state of matter, interatomic or intermolecular distances are at a minimum. However, this does not mean that solid matter is as solid as we perceive it with our sense organs. Almost the entire mass of an atom is concentrated in the tiny nucleus that holds the massive subatomic particles known as protons and neutrons. The electrons exist in various shells or orbitals outside the nucleus and are maintained in position by their electrostatic attraction to the protons in the nucleus. However, the mass of an electron is only about one two-thousandth that of a proton and the shells containing the electrons are many thousands of times greater in diameter than that of the nucleus. To put this in perspective, suppose the nucleus of a typical carbon atom was the size and weight of a grain of sand. The electrons would be comparable to dust particles some seventy feet from the nucleus. Clearly, an atom is almost entirely empty space! The only reason the empty space composing your hand does not immediately pass through the empty space composing a table is that the outermost shells of the atoms in your hand and those in the table are already filled with electrons. The electrostatic repulsion of the electrons in your hand with those from the table prevents your hand from passing right through atoms of the table. This creates the illusion of solidness, when in fact common matter is mostly just space.

Nonetheless, a solid object such as a rock or planet is ultimately composed of consciousness—albeit in its crudest form. This is the nadir point of the Creation Cycle. Consciousness completes the *saincara* or centrifugal phase of the cycle when solid material such as a rock or protoplanet is formed.

Saincara is the inanimate phase of creation and represents the formation of Cosmic Mind and the macroscopic universe. The process of *prati-saincara* commences at this point. It is the counter-movement or centripetal (center-seeking) movement of the Creation Cycle. In the *prati-saincara* phase of creation the unit or solid object begins to evolve mind. The same layers of Cosmic Mind form but in a reverse order, and they are constrained within physical boundaries of individual objects and living beings.

Transformation of matter into living organisms takes place gradually due to constant clash and the increasing reflection of Cosmic Consciousness (Cosmic Nucleus or *Purushottama*). As the proximity of the unit mind to *Purushottama* increases in this phase, the mind expands and progresses until the gradual expansion brings about similarity of the microcosm and the macrocosm and the unit finally merges into *Purushottama*. This completes the cycle. However, in order for a unit or microcosm to get scope for physical and mental development, life as we know it must first develop.

The *Prati-Saincara* Phase of Creation

Life can only develop on a planet when there is a proper balance of the five fundamental factors: ethereal, aerial, luminous, liquid, and solid. On a planet like Mercury or Venus, life is not possible since there is an over-abundance of luminous factor resulting in very high temperatures and no liquid water. On the other hand, frozen planets like Uranus and Neptune have insufficient luminous factor reaching their surface from the Sun and are too cold for life to evolve. On gas giants like Jupiter and Saturn there is an overabundance of the aerial factor and essentially no solid or liquid surface for life to evolve. At one time Mars may have been hospitable for the development of living organisms since there is strong evidence that it once had abundant and flowing water and a robust atmosphere. Further exploration of Mars may well prove that primitive life forms evolved there billions of years ago. However, today the lack of liquid water and a suitable atmosphere that would shield potential life forms from deadly solar radiation suggest that conditions are not now conducive to life.

In our solar system, that leaves Earth where life did evolve. Earth is believed to have formed about 4.5 billion years ago. Initially it was a hot molten body that was continually bombarded by meteorites and comets. The heavier elements such as iron and nickel settled into the core of the planet, where they are still found today in a molten state, kept warm by the radioactive decay of unstable heavy elements. The lighter substances gravitated to the surface of the planet and cooled, creating a hard crust.

Meteorites and comets containing liquid water and organic compounds including amino acids rained down on the surface of the infant Earth for hundreds of millions of years creating oceans and continents by 4.3 billion years ago. A thick atmosphere formed consisting mainly of carbon dioxide, water vapor, sulfur compounds, methane, and nitrogen. The atmosphere helped Earth retain its liquid water and shielded potential life forms from deadly solar radiation. The conditions found on the surface of the infant Earth were conducive to the formation of life.

Life was bound to evolve on the planet since whenever possible consciousness tries to express itself in the form of unit living beings, beginning with the formation of simple single-celled organisms. The oldest evidence of life on Earth is the fossilized remnants of cyanobacteria, dated at 3.5 billion years. These bacteria obtained energy via photosynthesis, which utilizes energy of the sun. A byproduct of photosynthesis is oxygen, and it is believed that these early single-celled organisms were responsible for converting the early atmosphere, which was devoid of oxygen, to one containing oxygen. The oxidizing atmosphere was deadly to many of the other microbes that were present on Earth at the time, but it was a vital component needed for the evolution of more complex organisms, including animal life.

In the absence of an organizing principle (*Purushottama*), the incredibly complex combination of atoms and molecules that makes up even the simplest life form would appear to be highly improbable. However, the Cosmic Entity relentlessly endeavors to express itself as unit consciousness in the *prati-saincara* phase of the Creation Cycle. Thus, life will develop on a celestial body whenever conditions permit it. Since the universe contains a countless number of stars, many of which have planetary systems, the universe must contain innumerable star systems that have extraterrestrial life.

Within our own Milky Way galaxy the fraction of stars having planets is small, and only a small fraction of planets have conditions conducive for the development of life. There would probably be only a small percentage of those planets where life developed that have had favorable conditions long enough for the evolution of highly intelligent living beings. This, along with the vast distance between stars within a galaxy, makes it unlikely that we will ever make physical contact with intelligent beings from another world.

The development of living organisms from inanimate matter through the action of the static force (*tamoguna*) on the manifest *Purusha* in the *prati-saincara* phase of creation involves both the actions of the vital energy, or prana, and microvita. A detailed discussion of microvita is beyond the scope of this book.[3] Nonetheless, life is an inherent property of Cosmic Mind, and it is apparent that it will manifest itself on a celestial body given the right conditions. With the first spark of life on a planet, the *prati-saincara* phase of creation begins in earnest.

Only unicellular organisms populated the earth for a very long time. Even single-celled life forms express unit consciousness and possess a very rudimentary unit mind. Naturally, they experience constant clash and struggle to survive. In this struggle for survival, the simple unicellular bodies found it advantageous to colonize with other cells and eventually multicellular organisms evolved. The time when multicellular organisms arose is difficult to determine with any precision but it may have been about one billion years ago. This was a giant evolutionary step; not surprisingly, it took a very long time before multicellular organisms began to dominate the biosphere.

Organisms evolve in the *prati-saincara* phase of creation through the constant struggle for survival and the constant pressure of the creative force. Struggle or clash causes a powdering down of the unit mind-stuff that leads to greater and greater expression of subtler *Citta*, *Ahamtattva*, and *Mahattattva* in the organism. Hence, organisms experience mental dilation, which is accompanied by greater physical complexity. Life forms gradually evolve into higher and higher species such as fungi, plants, invertebrates, vertebrates, and finally mammals.

This path toward the *Purushottama* is not without starts and stops as many life forms come and go when they fail to adapt to changing conditions. However, the movement from crude, less developed life forms to mentally subtler and thus more complex life forms is inevitable because like a cosmic clock the hands of evolution always move forward under the prevailing force of *Prakriti* and the attraction for the Great.

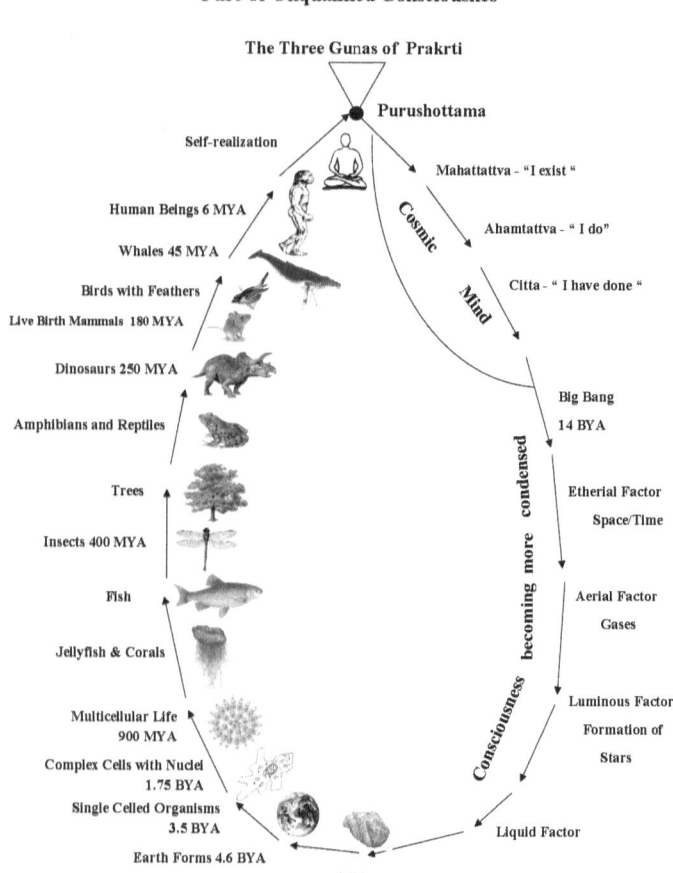

Brahma Chakra or Cycle of Creation

Darwin proposed that the evolution of species takes place because of natural selection or survival of the fittest. Small changes in the genetic material of an organism over a long period can lead to major changes in that organism or even the development of new species. Today biologists recognize that changes in the genetic makeup of an organism depend on changes in their

DNA. They believe that most of the changes in DNA involve random single-point mutations. However, random mutations and natural selection do not explain the complex systems found in living organisms.

Evolution of species must be teleological or goal oriented. The evidence indicates that evolution depends on the design of Cosmic Mind. The influence of Cosmic Mind is most apparent when we look at the development of complex structures, such as the blood clotting system in mammals. No less than eleven enzymes are involved in this very complex cascading system with several feedback loops. If any one of the enzymes is missing or defective it will normally be a death sentence for the organism. This is observed in hemophiliacs who are missing only a single enzyme. Such a complex system requires that all the components be present before there is any advantage conferred to the organism.[4] Hence, it is improbable that such a system could evolve in a step-wise manner via random mutations of DNA with selection of the fittest.

The materialists approach to evolution also fails to explain how mind and consciousness evolved. These are essential features of living organisms and any theory that fails to explain how they evolved must be considered fundamentally flawed.[5]

Development of Intellect

Once *Ahamtattva* begins to surpass *Citta* then intellect begins to develop in a living organism. In Sanskrit, the word for intellect is *buddhi*, which is related to Buddha, meaning the "Enlightened One." *Citta* is instinctive mind, but when more and more "I do" feeling, or *Aham*, develops in a creature, it begins to learn from its experiences and modify its behavior accordingly. For example, a dog learns a series of tricks through extensive training. This learned behavior involves some degree of *buddhi* since this behavior is not instinctual for the dog. Most animals show some degree of *buddhi* since they learn from experience and modify their behavior because of those experiences.

Higher animals such as apes, dolphins, whales, seals, etc. have a fairly well developed intellect or intelligence. They are able to solve problems using abstract reasoning. That is, they exhibit behavior that is not trial and error but seems to involve the ability to apply previous knowledge to a new situation.

Aham is created from *Citta* by psychic clash; that is, by stressful situations that force the individual's psychology to adapt and grow. In a sense, a portion of the *Citta* is powdered down or transformed into *Aham*. The very struggle for survival creates conditions that lead to this powdering down of the *Citta*. For example, a young deer wanders down to the river for a drink, but a crocodile is waiting submerged, hidden from sight. As the deer leans over for a drink of water the crocodile leaps toward it, jaws open. The deer escapes almost certain death when it jumps back just before the jaws of the crocodile are about to close on its leg. From this frightening experience, the deer learns to be more cautious when approaching the river for a drink of water.

Animals appear to have some degree of cognitive interconnectedness. A new learned behavior of one or more animals is passed along to other animals of the same species via Cosmic Mind. Because instinctual behavior is guided by Cosmic Mind (principally Cosmic *Citta*), learned behavior of individual animals can theoretically be passed on to others, helping to advance the evolutionary development of the species.

Another example of the connection animals have with the Cosmic Mind is the many stories of lost animals returning to their owners. Larry Dossey relates one such story in his book *Recovering the Soul*.[6] Bobbie, a female collie, was traveling with her family from Ohio to their new home in Oregon. During a rest stop in Indiana, Bobbie ran off and could not be found. After many hours of searching, the family gave up the search for Bobbie and continued on to Oregon. After about three months, Bobbie turned up at the doorstep of their new home in Oregon. Bobbie had never been to Oregon and there was no mistaking her because she still had the same collar (without ID tags) and other identifying marks. How could the dog find the new home of its owners unless it was connected mentally to the nonlocal Cosmic Mind?

There are hundreds of such stories involving cats, dogs, and birds. Within nature there are many more mysteries involving homing behavior. As noted in Chapter 3, scientists have hypothesized that such behavior depends on physical clues such as the position of the sun, stars, geomagnetism, and geographic features. However, the simple explanation for such behavior is that animals are guided by instincts arising from the nonlocal, omni-present Cosmic Mind.

Evolution of Man

The relentless march of evolution is slow. It has taken nearly a billion years for intelligent life to evolve on Earth from the first multicellular organisms. Homo sapiens (modern humans) evolved from hominid ancestors about 250,000 years ago. But how do human beings differ from other animals? As creatures develop mentally on the path of *prati-saincara* the unit *Citta* is increasingly transformed, giving rise to more and more *Ahamtattva* (I do) and *Mahattattva* (I am). The individual or unit mind grows in proportion to how much the Cosmic Mind is reflected in the mental plate of the organism. As the mind dilates, the physical structure becomes more complex with more and subtler glands in order to adjust to the higher psychic sentiments and demands. The ego, or sense of doership, develops followed by an increasing sense of self-awareness. Once the sense of "I exist" or *Mahattattva* becomes predominant, the organism is fully self-aware and self-determinant. We call such creatures human beings.

Humans have a developed sense of self-directed willfulness or ego. They can move their mind according to their desire and appear to possess free will. In plant and animal life, which lack the developed sense of "I exist" or unit *Mahattattva*, there is no ability to act independently. They act according to the guidance of Cosmic Mind. Hence, lower forms of life are guided by instincts and do not possess the ability to mentally move against the natural flow of the Brahma Chakra. Man, however, has the ability to point his mind in any direction he chooses, and mind always takes on the qualities of the object of its ideation. This is a double-edged sword since man can choose to focus on the subtlety of the *Purushottama* and move forward on the path of evolution at an accelerated pace, or he can choose to direct his mind toward the crude and move backward toward the unconsciousness of animal existence.

As organisms developed physically, they evolved sense organs in order to better interact and adapt to their physical surroundings. With the exception of *akasha*, which can only be sensed mentally, organs developed that allowed organisms to sense the other four fundamental factors (aerial, luminous, liquid, and solid). The five principal senses are smell, taste, sight, hearing, and touch. In Tantric philosophy, the physical stimuli that affect the sense organs are termed *tanmatras*, and the sense organs are called *indriyas*. The *indriyas* consist of the gateway organ (such as the nerves in the skin), the nerves connecting the gateway organ with the brain, and finally the portion of the brain where perception of the stimuli actually takes place. The seats for the five sensory organs are actually in the brain. For example, we actually "see" our outside environment on the visual cortex inside our brain. No light enters the brain, only nerve signals from the optic nerve. This is true for the other four senses as well. Finally, the *tanmatras*, or inferential waves, consist of a small portion of one of the fundamental factors. The ear (sound) and the sense of touch (wind) detect the *tanmatra* of the aerial factor. The luminous factor is detected with the eyes and by the sense of touch (heat), and for the liquid and solid factors one can add the senses of smell and taste.

Our mind is able to translate the nerve impulses reaching the brain from the gateway organs into externally projected reality with the help of our unit *Citta*. Similarly, when we experience thoughts and daydreams it is our unit *Citta* that is transformed in the process. Hence, the mind always takes on the properties of the object of its attention, whether internal or external. When the mind is involved with sensing and reacting to the external environment, or thinking about crude objects in that environment, it is difficult for it to expand and become subtler.

Humans, however, are dominated by unit *Mahattattva*, or "I feeling," and are thus naturally attracted to the *Purushottama*. The difference between this Cosmic Nucleus and the unit consciousness is that the unit beings are multi-purposive and unilateral while the *Purushottama* is multi-lateral and uni-purposive. Units are multi-purposive because they desire many things, but they are unilateral because they can do only one thing at a time. The *Purushottama* or Cosmic Nucleus is multi-lateral because he guides and witnesses everything in creation, but he is uni-purposive because he has only one desire—that all his creations return and merge with him. Hence, as hard as human beings may try to go against the subtle pull of the Cosmic Nucleus, the unit consciousness ultimately finds it impossible to go against the natural flow of the Cosmic Cycle that draws it nearer and nearer to its destiny of merging with him.

Spiritual Practice or Sadhana

The act of yoking one's mind to the subtle pull of the *Purushottama* is known in Sanskrit as sadhana. In English, this word is sometimes translated as spiritual or intuitional practice. Sadhana is an introversive or center-seeking movement of the mind. Many human activities fall under this category. In fact, any activity whereupon the ego is set aside and the mind is absorbed in the creative flow of invention, problem solving, artistic creation, or other such mental activities can be called sadhana. Such activity expands the mind and generates pleasure or euphoria. In practice, the most powerful types of sadhana are those where there is a willful turning of the mind inward toward the Cosmic "I feeling" and Cosmic Nucleus. An example of this type of sadhana is meditation on the Supreme Consciousness. Sadhana greatly accelerates the natural movement of the unit consciousness toward its goal of merging with the *Purushottama*. This accelerated movement of the unit toward its goal of merger is termed *vidya* in Sanskrit.

The opposite movement, termed *avidya*, is also possible for humans. If the mind is strongly attached to physical objects and desires like money, pleasure, name, fame, etc. then it takes on the qualities of these objects and desires and becomes increasingly crude. The movement from subtle to crude is *avidya*. It can also be an accelerated movement because human beings possess the ability to focus their attention as they please.

For animal life, such counter-movement or *avidya* is impossible. Lacking developed *Mahattattva*, they are guided by Cosmic Mind or instinct. They cannot go against the evolutionary flow of *prati-saincara*. This flow of evolution, however, is very slow. For example, it takes millions of years for organisms with higher intelligence like human beings to evolve from distant primate ancestors. This is why free will is a double-edged sword. It confers in humans the ability to practice sadhana and greatly accelerate the slow pace of evolution to attain merger with the Supreme Consciousness. However, at the same time free will makes it possible for the individual consciousness to regress rapidly on the evolutionary path and possibly return to animal existence.

PATH OF BLISS: THE FINAL JOURNEY OF MAN

The final merger of the unit consciousness with the Cosmic Consciousness or *Purushottama* is the final stop in the Brahma Chakra. This is a return to the starting point of creation. All units eventually finish their journey by becoming one with the Cosmic Nucleus. This occurs as the individual mind recognizes more and more completely that it is an inseparable part of the Cosmic Mind. In this merger of the unit with the Cosmic, the identity of the unit is lost—just like when a grain of salt falls into the ocean. The grain of salt merges with the ocean. Even those that may have regressed mentally to take on an animal existence will eventually move forward by the force of *Prakriti* to become human once again and then move on the path of *vidya* toward ultimate merger with the Supreme.

The path of knowledge or *vidya* is called ananda marga in Sanskrit, the path of divine bliss. When an individual finally realizes that their true mission in life is to try to merge with the Supreme Entity then they may decide to embark on the path leading to that goal. Spiritual practices or sadhana are like the footsteps that move one forward on the path. Paths vary according to the person, time, and place. Just as there can be many paths leading to a mountain peak, so there are myriad roads or disciplines leading to the realization of the Supreme. However, the path taken will not be without difficulties.

Spiritual practices break down the ego or *Aham*. This can be both difficult and painful, and often one's growth is measured more by the difficulties experienced and overcome than by the pleasures one feels. Once one embarks on the path of *vidya*, however, the mind begins expanding more rapidly and becomes increasingly subtle. There is an accelerated movement of the unit mind toward the ultimate goal. It is like a small fish living in the stagnant waters of a pond that leaps a dam into a tiny brook. At first, the flow is a trickle but soon it increases exponentially to that of a great river, and eventually that river reaches the ocean. Hence, the path is also blissful since the individual feels that they are finally moving toward their goal and the experience is one of greater and greater happiness as they draw closer to that goal.

The Purpose of Life

Obviously, there is purposefulness at work in the Brahma Chakra. The Supreme Witnessing Entity or *Paramapurusha* is transformed under the influence of the qualifying force, *Prakriti*, into myriad forms and beings that are inexorably drawn back to the source that created them. If we use the analogy of a clock then we human beings are at the 11:59 point in this cycle. Whether we consciously realize it or not, we are being constantly pulled toward finishing the cycle and merging our little "I am" into the Cosmic "I" and into the unqualified Consciousness of the Cosmic Nucleus.

This is the true purpose and meaning of our existence; to deny this purpose is to throw ourselves backward toward the darkness of unconsciousness. However, the question remains, why did the unqualified *Purusha* allow himself to come under the bondage of *Prakriti* in the first place? There is no way to answer this question. Some call it his *liila* or play. They say he was bored because there was no play to witness. Was he really? We will never know the answer to this question without becoming one with him.

OTHER THEORIES OF CREATION

All scientific theories of creation rest on the assumption that something was present at the time of creation. Whether this something was a singularity with the potential of almost infinite mass and energy or just quantum particles, there is always the problem of explaining where the stuff of the universe came from and how it became so organized. That is, what is the first cause of creation?

Most cosmologists or scientists that study the origins and nature of the universe try to avoid such questions by saying that these ultimate questions are the purview of metaphysics or religion, not astrophysics. More and more scientists, however, have turned to the study of consciousness as an explanation for the origin of the universe. The Brahma Chakra theory of creation provides an explanation for how the universe and life originated from Consciousness. It answers the ultimate questions in an elegant way, and it refrains from the illogical proposition that the first cause or God is a separate entity from his creation.

5
The Rule of Three

In the previous chapter, we discussed how the Cosmic Binding Principle, *Prakriti*, has three different binding modes: sentient or *sattvaguna*, mutative or *rajoguna*, and static or *tamoguna*. The sentient binding mode produces the subtle "I feeling," while the mutative force converts some of the "I feeling" to "I do." However, it is the static force of *Prakriti* that does the heavy lifting. *Tamoguna* is responsible for converting some of the "I do," or *Ahamtattva*, to Cosmic Mind-stuff, and from there through further binding, it produces the five fundamental factors, leading eventually to the solid factor.

The transformation of Consciousness into different forms is somewhat analogous to the three forms of water: vapor, liquid, and solid. To the unschooled eye it would not be obvious that these three entities are composed of the same stuff—common H_2O. Interestingly, the three binding forces that shape the Cosmic Mind and transform it into the physical world are reflected in many of the mental and physical forces and properties of the universe.

The Three Forces of the Physical Realm

Until the early 1970s, physicists described four fundamental interactions in nature. The strong nuclear force, which binds protons and neutrons together in the nucleus of the atom; the weak nuclear force, which is responsible for radioactive decay; electromagnetic force, responsible for binding charged particles and for all types of radiation that travel at the speed of light; and finally gravity—the weak attraction between bodies possessing mass.

Scientists have now demonstrated that the weak nuclear and electromagnetic forces are not truly separate forces and the combined forces are called the electroweak force. Therefore, there are only three fundamental forces at work in the physical universe, and all of the known interactions—chemical, electrostatic, nuclear, atomic, and mechanical—are manifestations of one or more of these three basic forces.

The distances covered and strengths of the three forces differ greatly. The strong nuclear force is extremely strong since it must overcome the mutual repulsion of protons in the nucleus of atoms; however, it only acts over an extremely short distance. The electroweak force is intermediate in strength and in atoms, it acts over a short distance, but as electromagnetic radiation, it can cover great distances. Gravity is an extremely weak force but it can act over astronomical distances. The three forces correspond to a subtle force—gravity, a mutative force—electroweak, and static force—strong interaction.

Our current understanding of these forces is that they act at a distance and are therefore field forces. That means that the greater the distance between objects the weaker the force. For gravity and the electrostatic attraction between a positively charged particle like a proton and a negatively charged particle like an electron, the force diminishes by the square of the distance between the particles. For both the strong nuclear force and the electroweak force acting within the nucleus of an atom, the force field is more constrained and it is complicated to describe mathematically. Quantum field theory utilizes intermediate particles (gauge bosons) as carriers of the forces. For example, the electromagnetic attraction between an electron and proton is carried by the photon, while the strong nuclear force is carried by the gluon and the gravitational force by the graviton.[1]

The theory combining electromagnetic and weak nuclear forces (electroweak theory) when combined with the theory describing the strong nuclear interaction form the so-called Standard Model of particle physics. This Standard Model has been very successful in predicting and describing with great accuracy experimental results over an enormous range of energies. The Standard Model has been hugely successful in providing a fundamental understanding of the interactions and composition of matter and energy at the quantum level, but it has not been able to explain how certain heavy particles exist.

In order to complete the Standard Model, physicists hypothesized the existence of a yet unseen particle called the Higgs boson. This particle is sometimes referred to as the "God Particle" after the title of a book by Leon Lederman: *The God Particle*.[2] The Large Hadron Collider at Cern has now provided conclusive experimental evidence for the existence of the Higgs boson. Its existence completes the Standard Model and explains how particles having mass formed during the Big Bang.

The feeble force of gravity is the force most responsible for shaping the universe. The matter and energy created by the Big Bang was not perfectly distributed or homogeneous, as evidenced by the nonhomogeneity of the microwave background radiation. The lumpiness of the primordial gas (mostly hydrogen) created the condition needed by gravity to pull slowly the gas together to form stars and galaxies. If the gas had been distributed evenly in the initial phase of creation then each particle of gas would have been subjected to an equal pull in all directions from neighboring particles and no clumping or aggregation of particles would have taken place. However, such a scenario seems improbable in such a gigantic and chaotic explosion as the Big Bang. Any dissymmetry in the original distribution of gas would inevitably lead to the formation of clumps that could grow in mass by the weak pull of gravity and eventually become massive, dense, and extremely hot resulting in the formation of stars.

Gravity is a very weak force until an object becomes very massive. For example, the force of attraction between two ball bearings is so small that it is immeasurable. However, the gravitational attraction between the earth, with its great mass, and a ball bearing is sufficient to accelerate the ball bearing rapidly to the ground when dropped from a height. Isaac Newton was the first scientist to describe mathematically the force of gravity between two objects. For a massive object like our Sun, the gravitation force of attraction is very large and extends far into space keeping the eight planets in orbit around it as well as millions of more distant objects. For supermassive objects such as black holes the force of gravity is so extreme that even light cannot escape its grip.

Thus, the subtlest force, gravity, has done more to shape the universe than the mutative electroweak force or the static strong nuclear interaction, and under cosmological conditions, it acts at a much greater distance and with much greater power than the other two forces of nature. This goes along with one of the fundamental characteristics of the Unity Principle, namely that real power lies in the subtle, not the crude.

As mentioned earlier the three fundamental forces appear to be very finely tuned in such a way as to allow for the formation of stars, planets, and organic life. If any one of the three interactive forces differed by even a few parts per thousand it probably would have changed the nature of our universe profoundly, making life as we know it impossible. For example, the strong nuclear force keeps protons in the nucleus of atoms glued together. However, this force is just the right strength to allow some heavy atoms to disintegrate spontaneously in a process called radioactive decay. In the absence of radioactive decay, with its consequential release of energy, Earth would not currently have a molten core and be active geologically or have a magnetic field shielding living organisms from deadly cosmic radiation. Hence, life might never have evolved on the planet.

The strength of the strong nuclear force is also just right for the fusion of hydrogen nuclei to form helium along with the release of significant energy. It is chiefly this reaction that powers the stars. At the center of a star, gravity and temperatures are just right to form heavier elements from the fusion of helium and other nuclei formed from fusion. These reactions also give off energy, both prolonging the life of stars and providing the building blocks for life. Finally, when iron forms, fusion ceases to give off energy and iron builds up in the core of a star as it slowly runs out of fuel and begins to die.

Hence, it seems that both gravitational force and the forces and energies involved with nuclear fusion are perfectly balanced to allow stars to form and maintain a constant output of energy for billions of years. Our Sun is approximately 4.6 billion years old and we can expect it to remain hospitable for another five billion years, after which it will begin to run out of fuel, initiating a phase of rapid expansion to a red giant that will eventually engulf Earth. Life on Earth has been around for over three billion years and may have another five billion to go before Earth is burnt to a crisp by our expanding Sun. A long-stable sun is probably important for the evolution of intelligent life on a planet, and it is unlikely that this would occur without the fine-tuning of the fundamental forces of the universe.

The electrostatic component of the electroweak interaction is also perfectly tuned for the evolution of life. For example, the precise strength of attraction between protons in the nucleus of atoms and electrons in their shells has led to the formation of stable chemical compounds needed for the evolution of organic life. Fortunately, for us these forces allow hydrogen and oxygen to form a stable compound called water.

The properties of water are quite different from those one might expect based on analogous compounds. For example, hydrogen sulfide (H_2S) is a similar but heavier compound than water, yet it is a gas at room temperature. The unique properties of water, such as its very high freezing and boiling points, solvating power, and expansion upon freezing are all due to its strong polarity or asymmetric charge distribution. These properties have been essential for the evolution of life on this planet. If the universal electromagnetic forces at work holding molecules together and the similar forces responsible for creating polarity within molecules were even slightly different, then it is unlikely that a substance such as water, with the unique properties needed to support life, would exist.

Even assuming the three forces shaping the universe just happened by chance to have exactly the right strengths and properties to allow for the formation of stars, planets, water, and organic compounds, chance alone does not offer a satisfying explanation for the origin of life on our planet. If we assume that primordial life was RNA or similarly based, that its replication only required one enzyme (protein); and that the earliest life form had many more than twenty-one amino acids available for making proteins, the odds that a self-replicating life form could develop solely by chance are extremely tiny. Many authors have tried to calculate the odds that life developed spontaneously by chance and concluded that it is a statistical impossibility. However, since we have no way of ascertaining the complexity of the earliest life form, such attempts are futile.

As pointed out in Chapter 3, the seemingly perfect balance of forces needed for the evolution of life in the universe has been termed the Goldilocks enigma. The Unity Principle requires that the perfect balance among the three physical forces that make the universe suitable for life, and the combination of molecules needed to make a complex self-replicating life form, are not chance occurrences. Everything in the universe comes within the purview of Cosmic Mind and is therefore the product of conscious design. Nothing falls outside the scope of Cosmic Consciousness and hence nothing occurs by random chance. When we look out into the vastness of space and see the wonderful beauty and complexity of the physical world and the exquisite dance of life on this planet it suggests a Great Designer.

Suppose we travel in a space ship to a distant world. Walking outside our spaceship, we find a hand-held computer. Intuitively we would know by looking at the complexity of the device that it was manufactured by some intelligent being and did not come about merely by chance. The life we witness on our planet is millions of times more complex than a computer. The simple explanation for its creation is that it is the product of an unseen intelligence.

The Three Principal Layers of Mind

Our body and conscious mind are the crudest layers of our existence and are dominated by the static principle. The conscious mind is mostly involved with the physical body and controls the motor organs. It also receives input from the sense organs in the brain. The needs of the physical body come first and it's not until those needs have been satisfied that individuals are able to devote the time and energy necessary to discover the truth that they are more than this physical body.

The mutative principle of *Prakriti* dominates the subconscious mind. This mind is active all the time, but we are most aware of it when we are dreaming. The subconscious mind does a lot of its work nonverbally—that is, using images instead of words. Being subtler than the conscious mind it is more expansive and is involved with memory, comprehension, imagination, and deep thought. It is active at all times except during deep sleep, sorting memories and creating mental pictures that we are usually not conscious of. It is an untiring slave to the conscious mind, working constantly behind the scenes to solve problems and prepare us for success. It is the layer of mind that performs the higher mental functions, including philosophical thought, scientific reasoning, and information management. Lying between the conscious and unconscious layers of mind, intuitive ideas and solutions to problems often pass through it and emerge into consciousness during the quiet period just before sleep or during meditation.

The subtle layer of mind dominated by the sentient principle is the unconscious or superconscious mind. According to Tantra, the unconscious mind consists of three separate layers, as discussed in Chapter 7: the *atimanasa, vijinanamaya*, and *hiranmaya koshas*. The unconscious mind is extra-cerebral. It is an omniscient collective mind shared by all humans. It is the active reflection of the Cosmic Mind on the unit mental plate.

Beyond the unconscious mind lies the self or atman, which is the pure witnessing entity of our being. Sometimes this immortal self is called the soul. As we move from the cruder layers of our being to the subtle unconscious mind, we experience expanded awareness, knowledge, and energy. It is rare for an individual to penetrate deeper than their subconscious mind, since typically the conscious and subconscious minds are in a constant state of agitation, concerned with everyday problems and needs. It is only when the mind is quiet and at peace that it can experience the subtle, unlimited knowledge of the unconscious realm.

Occasionally a person may penetrate through the turbulence of the conscious and subconscious layers and experience for a moment the uninhibited flow of ecstasy, super-mundane knowledge, and unconditional love of the unconscious mind. After a while, the unrelenting attraction for the physical world draws them back to "normal consciousness." However, this firsthand mystical experience will forever change the way they view the world.

The Three Phases of Life

A baby is born innocent. It has no ego, no individuality, and no discrimination between self and non-self. In a way, the baby is like an animal guided only by instinct or Cosmic Mind, oblivious to problems and attachments, totally accepting and trusting of its world. It knows only love. It lives totally in the timelessness of the present moment. This state of pure, unconditional being is similar to the state of consciousness that we strive to attain in life. However, this perfect state of bliss and contentment does not last, nor is the child without an inborn history. It is not born as a "blank slate." Each child brings with it a different set of samskaras—traits that give the child its character and individuality. The mother instantly recognizes her baby from all others and knows its peculiarities. So even very early on in a child's development, remnants from their previous life on earth begin to color their individuality and self-image. However, their consciousness of self is undeveloped along with their memory, which for a young child is non-continuous or sporadic. Only a few memories are retained prior to the third year, to say nothing of a previous life, but samskaras differ from memories in that they originate from a deep, unconscious layer of mind and directly affect the physical and psychic bodies.

Gradually as needs and desires arise that are not immediately fulfilled, pain arises, and the beginnings of fear, anger, desire, attachment, and doubt emerge. In order to cope with the physical world the child gradually learns the difference between self and non-self and begins to speak of itself in the first person. The developing ego imparts a sense of duality—the difference between "me and it" or "me and you." The ego separates what is physically real from what is unreal. It helps the child organize its thoughts and make sense of them and the external world. This development is essential for the healthy growth of the child, for without it the discrimination and judgment needed for survival would be lacking.

As the child develops mentally and physically it develops empathy, learns social rules of behavior, and forms friendships. However, the child's character is not solely a product of their nurturing. A good portion is inherited. Inexplicably the child may take a life path that is unexpected and very different from that of its parents. This is because the individuality of a child is dependent on both the parental family and environment as well as upon the samskaras it inherited from previous lives, which need expression in this life.

Ego development continues into and beyond adolescence as strong emotions of sexuality, love, dislike, and anger take hold. Ego development does not end here, however. People believe that happiness comes from being an accomplished human being. To be happy one needs a meaningful and well-paying job, a spouse, children, a nice home, and personal conquests. The ego's quest for achievement has no real boundaries because underneath the "I do" mentality of ego lies the limitlessness of the "I am" realm. However, as the ego amasses layer upon layer of possessions, wealth, status, and power in its search for happiness, its burden grows. Possessions are lost or lose their attractiveness over time. A lack of fulfillment and a sense of dissatisfaction with life may develop.

Old age creeps up and the motor organs become less adept, the sense organs less keen. Extroversive activities do not bring as much pleasure as in youth. Trapped in the space-time of "me" and "mine," the ego fails to experience the true happiness that lies beyond the bondages of time, place, and person. The static binding force of *Prakriti* dominates ego development. While it is essential for the first phase of human development, the vast majority of people remain trapped by its selfish allure and do not graduate to the next level—that of the seeker.

A seeker is someone who has entered the *vidya* path of self-discovery. A seeker has acquired wisdom and realizes that knowledge of self is knowledge of God. Such a person rejects the idea that extroversive activities bring lasting happiness. They have learned that true happiness is born from within. A seeker is someone who has begun the search for the meaning of life. They have crossed the threshold from the stagnant waters of the pond into the flow of a stream that will lead eventually to the ocean.

Seekers are characterized not only by an interest in spirituality but also by the practice of some form of introspection such as meditation. In their search for truth such individuals become tolerant, nonjudgmental, loving, selfless, service-minded, more in tune with their body and nature, healthier, and happier than they ever knew possible. Not every step on their path is blissful since they are still encumbered with negative samskaras from the past and plenty of ego baggage to overcome, but the rewards they feel along the way offer convincing proof that they are on the right path.

Deepak Chopra describes the inward marks of the seeker as: giving, motivated by selfless love and compassion, wanting nothing in return, not even gratitude; intuition becomes a trustworthy guide to action, replacing strict rationality; one catches glimpses of an unseen world as the higher reality; intimations of God and immortality appear. These signs will be accompanied by growing enjoyment of solitude, by self-reliance in place of social approval, by stirrings of being, and by willingness to trust.[3]

To graduate from ego attainment to seeking union with the Infinite is a normal and healthy change that takes place as we age. However, for many people the transition never takes place. They remain engrossed in the crude pleasures of the body and physical world. They fail to learn of the untold benefits gained by powdering down the ego and searching within for peace and happiness and the answers to life's mysteries. In his book *Modern Man in Search of a Soul*,[4] Carl Jung described the problems such people encounter as they age. They try to cling to the pleasures of youth as they enter the second half of life. They are accustomed to acquiring most of their pleasures in the external world, but their sense organs inevitably become duller and the motor organs weaker. Such people seek greater and greater stimuli in order to compensate and may fall prey to what is called a "mid-life crisis."

If left uncorrected, this attempt to stem the tide of old age can lead to an autumn of life marked by discontent, dissatisfaction, cynicism, and unhappiness. No matter how hard one tries, it is impossible to reverse the physical effects of aging. Jung argued that for the aging person it is a duty and a necessity to give serious attention to their personal psycho-spiritual development. The individual is freed from many of the mundane obligations of youth in their middle years, and they have the opportunity to reap the incredible treasures that an introspective approach to life can provide—for true happiness springs from within, not from without.

Some other terms used to describe a seeker are sadhaka, spiritual aspirant, chela, adept, yogi, Tantric, monk, nun, acharya, disciple, devotee, and spiritualist. A seeker may endeavor to overcome ego or *Aham* but is still largely under its influence. The mutative binding principle of *Prakriti* dominates in the seeker.

The most advanced or subtlest phase of life is that of the seer or saint. The seer sees everything as a manifestation of the One. For such an individual the illusion of separateness breaks down. For the seer, unity is not something taken on faith or a belief—it is experienced. The seer has surrendered the ego and experiences the world in its pure form—as Consciousness. Hence, the seer feels at one with every living creature and with every particle composing the universe.

The seer is completely open, no longer plays psychological games, and is incapable of feeling any emotion except unconditional love for God and all his creation. The seer lives totally in the now and has access to the unlimited knowledge that lies in the collective unconscious and the Cosmic Mind. Seers are unattached to the fruit of their actions; they feel that they are merely instruments for the will of the Supreme Entity. Thus, a seer is free, living in a state of grace and indescribable bliss.

It is natural for people to be drawn to such individuals and look up to them for advice and spiritual guidance. The seer is sometimes called a saint, spiritual teacher, guru, rishi, sadhu, master, sage, and prophet. The sentient force of *Prakriti* dominates this final phase of human development. Knowingly or unknowingly all human beings crave the limitlessness that accompanies this, the ultimate phase of life.

Food

Every object of the world is dominated by one of the three modes of *Prakriti*—sentient, mutative, and static. Food is no exception, and according to its intrinsic nature, it is divided into the same three categories. Foods that produce a subtle mind and body and are thus conducive to physical and mental well-being are sentient. Examples of sentient foods are grains, vegetables, fruits, seeds, milk, and milk products.

Foods that are good for the body and may or may not be good for the mind, but which are not harmful for the mind, are mutative. Examples are tea, coffee, caffeinated sodas, herbs, spices, nutritional supplements, and medicines.

Static foods are bad for the mind and may or may not be good for the body. Examples are meat, fish, shellfish, eggs, intoxicants, rotten food, mushrooms, fungi, onion, and garlic.

In order to have a balanced mind capable of quiet concentration or meditation a person needs to pay attention to the qualities of the food they eat. Static foods disturb and agitate the mind, or can cause inertia or sleepiness. Either makes it difficult to practice meditation. Mutative foods can stimulate the mind and make it possible to meditate without falling asleep. Examples are tea and coffee, which are mild stimulants, but lack nutritional value and are best consumed in moderation.

A diet consisting primarily of sentient foods with some mutative foods can be termed a meditation diet. It helps the sadhaka concentrate their mind and thus progress more rapidly on the *vidya* path. Such a diet is not strictly lacto-vegetarian since a few vegetables such as onion, garlic, and fungi may be considered vegetarian but are known to disturb the mind. The qualities of food also change with different times, places, and persons. For example, in a cold climate such as that of the Inuit people, animal flesh takes on a mutative quality. In addition, for some people a normally sentient food such as peanut may be harmful and hence static because they are allergic to it.

Everything Else

Because it is disorganized, *Prakriti* is impotent in the unmanifested Supreme Consciousness, which is called *Nirguna* Brahma. In Sanskrit, *nir* means "without" and *guna* means "binding." *Prakriti* holds sway over everything in the created universe or *Saguna* Brahma (*sa* means "with"). Since *Prakriti* has three modes, every person, place, and thing in the universe can be regarded as being dominated by *sattva-*, *rajo-*, or *tamoguna*. Naturally, seekers on the *vidya* path of bliss are drawn to persons, places, and things that have sentient nature as opposed to those with static nature. Thus, they seek the company of spiritually minded persons, enjoy music that elevates the mind, and enjoy locations that are peaceful and have expansive views and natural beauty. They tend to avoid the pitfall of over-consumption and tend to be happy while owning fewer things. The more one owns the more time one spends worrying about keeping and maintaining one's possessions. Hence, most seekers follow the adage that if you make a list of your possessions then its length is a measure of how far you are from God.

6

The Law of Action and Reaction

Newton's third law describes the law of action and reaction for physical bodies. It states that for every action, there is always an equal and opposite reaction; that is, the forces of two bodies acting on each other are always equal and are directed in opposite directions. In the physical realm, the action and the reaction occur simultaneously. A similar law applies in the mental realm. Sometimes it is called the Law of Karma. However, there is one major difference between action and reaction in the mental sphere—the reaction may be stored in the mind and experienced later.

We have seen how every physical action, whether of the motor organs or sense organs, involves Mind-stuff or *Citta*. This is also true for mental activity. The unit *Citta* is therefore involved with every conscious mental and physical action performed by the unit mind. When the *Citta* is vibrated, it leaves a permanent impression on the unit mind, including memory, and an associated reactive component called in Sanskrit samskara. The samskara is the stored reactive momentum or potential for future action. When the samskara is expressed, it is said to be burned or exhausted.

In a sense the burning of the samskara returns the mind to a more serene or perfect state. This process is analogous to forming a dent in a hollow rubber ball with your finger. The depression may last for a while, but if the ball is massaged with the hand, the dent can pop out creating a perfectly symmetrical ball once again. In this example, the dent represents the samskara. When the dent pops out the samskara is exhausted.

A samskara is called a reactive momentum because it is the reaction resulting from some action performed by the unit. The momentum in this case is stored as potential mental energy in the *Citta* of the unit mind. The expression of a samskara is accompanied by the release of kinetic mental or physical energy equivalent to the mental energy or impression that created it. Although the amount of mental vibrational energy may be the same, the type or quality expressed will normally be different. For example, a person does a kind deed for a stranger on a road, such as helping them to change a flat tire. The reaction to such kindness will be stored in their mind and perhaps come back to them later in the form of help someone else provides them. However, it probably will not be help changing a flat tire. In a subtle, often unconscious way, the reaction teaches the individual to behave nicely to other people.

Similarly, a bad deed will ultimately bounce back creating pain, unhappiness, suffering, etc. Such suffering creates conflict in the mind that powders it down. Unconsciously the individual learns that the action that created the samskara was bad, and they are less apt to repeat that action. This is essentially what is called the Law of Karma: we reap what we sow; or what goes around comes around. It is the basic ethical rule that most people live by. It is the basis for the "Golden Rule" to do unto others, as you would have them do unto you. This law of action and reaction is a reward/punishment system by which we learn to be better individuals, develop mentally, and ultimately gain wisdom.

Humans possess free will and may choose to act according to their own whims. Therefore, we are free to create samskaras, both good and bad, but we are not free as to when and how they are expressed. The greater the mental vibration or intensity of an action, the more powerful will be the associated samskara that is created and the stronger will be its effect when the samskara is expressed. Strong desires create a strong potential for action. For example, a boy goes to an air show and is enthralled by the aerobatics. He develops the desire to perform aerobatics himself. He may decide to go to school to become an aeronautical engineer, work hard at his job so he can afford to train as a pilot and buy a used airplane, and then spend many years training in aerobatics. It may be many years and many hundreds of hours in his aerobatic airplane before this strong desire burns off and he feels at peace not flying any more.

Actions performed unconsciously or with little or no thought do not create samskaras. Samskaras only arise when there is *Aham* or "I do" feeling (ego) involved. Therefore, the action of others, actions performed unconsciously, and acts of God, such as a flood do not create samskaras in our mind directly. However, we may still experience reactions to such things that create personal samskaras. The best way to perform actions without creating new samskaras is to perform the action with the ideation that we are not the actor. It is the Supreme Consciousness acting through us. In other words, to feel that we are a machine and he is the machine operator. In Sanskrit, the term for such egoless action is *madhuvidya*, which means "knowledgeable action." When the ego is not involved in an action the reactive momenta for the action remains in the Cosmic Mind and not in the mind of the unit.

Sadhana and Samskaras

Sadhana is the main instrument that advances one on the path of *vidya*. It involves focusing on the *Mahattattva* or "I am" layer of mind instead of the *Aham* or ego. This unit "I feeling" is a reflection of the Cosmic "I feeling" in the individual mind. Thus, the unit "I feeling" is actually nonlocal and universal. For example, imagine a radio station broadcasting music from the top of a mountain. The radio waves propagate in all directions and are received by the radio in your car. Many other cars also tune into the station, and although the cars all move around each picks up the same music.

Similarly, there is actually only one "I feeling" in the cosmos, the Cosmic *Mahattattva*. Our unit minds act like radio receivers and pick up the vibration of the Cosmic *Mahattattva* and it is translated by the unit mind into "I am Joe" or "I am Rebecca," etc. If through the intense concentration of sadhana we experience our own, unqualified "I am" feeling then we will be aware of how it transcends our little sense of self. We will know that this same "I feeling" is experienced by every living creature. This experience is sometimes called self-realization or *savikalpa* samadhi. The feeling is "I am one with God."

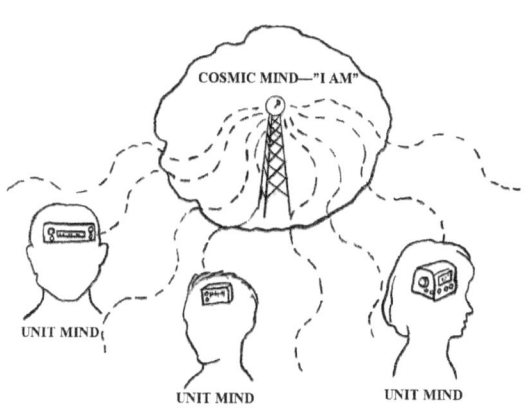

Similar to a radio station, there is only one cosmic "I am" broadcast throughout the universe that is picked up by all the unit minds.

Vidya or spiritual knowledge is the accelerated path toward the Supreme Entity. Samskaras are the chief impediments on this path, and sadhana is the practical method for quickly exhausting or burning these obstacles that impede one's progress on the path. The act of disassociating the mind from the physical world and focusing it on the Cosmic "I feeling" or Cosmic Consciousness has the effect of ripening many of the long-buried samskaras. These samskaras will then be experienced or "burned" by the spiritual aspirant. Sometimes this is experienced as rapid changes in life, feeling high, or feeling low. By practicing sadhana, spiritual aspirants can burn samskaras rapidly and by practicing *madhuvidya* they can refrain from creating new ones.

SAMSKARAS AND DEATH

Most human beings accumulate many samskaras during their lifetime. Many unfulfilled desires and stored reactions will not have been exhausted at the time of death. What happens to these samskaras when the mind and body separate at the time of death? Since consciousness is neither created nor destroyed, and the unit mind has within it innumerable vibrations in the form of samskaras, there must be some expression of the unit following physical death in order that these samskaras can gain expression. This is the only way in which the unit can complete its journey and merge with the Cosmic Entity.

Since many of our samskaras are created in the physical world with the physical body, it is reasonable to expect that the unit mind will need to find a new physical body in order to express the unexhausted samskaras from the previous life or lives. Under the guidance of Cosmic Mind, the bodiless mind is guided to a womb where it can best express its unburned desires or samskaras. The rebirth of the unit into a new physical body is called "reincarnation." Lord Buddha called this process the "wheel of birth and death." This wheel does not turn endlessly because eventually the unit develops wisdom and decides to embark on the path of bliss or *vidya*. We will discuss reincarnation in more detail in Chapter 9, but the following couple of stories illustrate how some children seem to remember their previous lives.

Beginning at age two, James, a native of New York had recurrent nightmares of being shot down and being unable to escape the burning wreckage of his P31 during the war. As memories of this previous life surfaced, he recounted to his parents some of those details, such as the name of the aircraft carrier he took off from, the name of a friend he flew with, and his own name. These details checked out with actual historical records. By age eleven, James was no longer bothered by remembrances of his past life and death.

Sita, age five, grew up in a small town in Bihar, India. She had a vivid recollection of a previous life as a mother with three siblings living in another town in India. She recalled being sick with pneumonia and saying farewell to her relatives as she lay dying in bed. Her parents were Hindu and believed in reincarnation. They were intrigued by her stories and decided to take her to the village where she claimed to have lived in her previous life. When they arrived by train in the village, Sita recognized the town and knew the way to the house she believed she lived in before. There she recognized two of her daughters who had grown up. She knew their names and even their pet names when they were children. She also knew many details about the family, including how one daughter got a scar on her neck when she was a young child. The children were convinced that Sita was indeed their mother in her previous life. To this day, the two families are very close.

7

The Nature of Unit Consciousness and Unit Mind

The *Paramapurusha* or Supreme Consciousness witnesses his creation without being attached or influenced by it. He has set the plan and *Prakriti* carries the plan out. He may be the efficient cause of creation and the guidepost, but the *Paramapurusha* performs no actions himself. The creation is internal psychic to the *Paramapurusha*. That is, everything in creation takes place within his mind. Thus, he is said to be omniscient or all-seeing.

This witness-ship extends to the unit level. When a person perceives an object with a sense organ the unit *Citta* is vibrated according to the nerve impulses reaching the brain. The same occurs when a thought or visualization takes place in the brain. Physical nerve impulses occur in the brain and are picked up sympathetically by the *Citta*. This interaction between physical brain activity and the purely mental stuff of *Citta* takes place at the quantum level and is therefore very difficult to demonstrate scientifically.[1] Nonetheless, our perception of the external world can take place only because the objective portion of our mind, the *Citta*, takes on the form of our perception. In order to do this it requires working sense organs in the brain and at the gateways to those organs. Hence, damage to the brain or to the gateway organ such as the eyes can cause blindness. Without brain activity, the *Citta* cannot function and there is no perception.

However, perception depends also on the ego or *Ahamtattva* to look at the *Citta* and witness it. Without "I feeling" or unit *Mahattattva* there also cannot be any sense of "I watch" or witness-ship of the *Citta*. Finally, we intuitively know of our own existence; there is a part of us that witnesses the *Mahattattva* or sense of self. This witnessing entity of the unit mind is the unit consciousness or atman. This atman is also called the *anu-purusha* or unit consciousness. It is identical in quality to the *Paramapurusha* or Supreme Consciousness. In quantity, it can be considered much smaller than the Supreme Consciousness since the atman is associated only with the unit mind.

In the same way that a microscopic portion of the sun's total radiation falls on our face when we turn our head toward it, only a tiny portion of the Supreme Consciousness is associated with the unit mind and witnessing it. Hence, the atman is the witness of all the layers of the mind including the body, and because it is nothing but a small part of Supreme Consciousness, the Supreme Consciousness or *Paramapurusha* witnesses everything that happens to and by the unit.

The unit consciousness or atman may be compared to a mirror. It is unchanging but reflects the dancing vibrations of creation. If the unit mind is withdrawn (transcended), so that all these vibrations cease and there is not even the sense of "I exist" or *Mahattattva*, then only the pure consciousness of the atman will be experienced. In this state, the unit consciousness has no boundaries and the unit atman merges with or is identical to the Paramatman or Supreme Witness.

The Layers of Unit Mind

Just as the Cosmic Mind has layers (*Mahat, Aham,* and *Citta*), the unit mind is also layered. The crudest, outer layer of unit mind is termed the *annamaya kosha* in Sanskrit. Here *annamaya* refers to the physical body created by food and *kosha* means "layer of mind." This *kosha*, the body, is composed of the five fundamental factors and is dominated by the *tamoguna* of *Prakriti*. Most of the functions of the *annamaya kosha* occur independently of our conscious thoughts. For example, we do not have to think about breathing, heart beating, or releasing digestive enzymes from the pancreas. Most of our glands function autonomically. These unconscious bodily activities are governed by the rudimentary mind of the body, sometimes called the autonomic nervous system.

The next layer of mind is the conscious mind or *kamamaya kosha*. This layer of mind is active when we are awake. It is home to our desires, fears, and pleasures. It is the crude mind because it is involved with sensing and acting in the external world of crude matter. The conscious mind governs the sense and motor organs, and it deals with bodily needs such as eating, drinking, sleeping, and procreating. It experiences fear and is involved with self-preservation. Man shares with animals the basic needs and qualities controlled by the conscious mind. The only difference is that animals are incapable of performing higher mental activities such as contemplation of the Supreme Consciousness.

The next subtler layer of mind is the subconscious mind (sometimes called the preconscious mind). This layer of mind is called the *manomaya kosha* in Sanskrit. *Manomaya kosha* literally means "the mental layer." This mind is active in the waking state and during sleep. Dreams are a function of the subconscious mind. Most memories lie in this layer of mind. When activated, memories are expressed in the conscious layer.

Memory is a complex process. Memories normally consist of exquisite details and mental pictures that we are not even conscious of most of the time. Sometimes these detailed recollections can be brought out by electrical stimulation of the brain or by hypnosis. For example, a person may recall under hypnosis a detailed description of a criminal perpetrator including details of his tattoo, earring, scar, etc. that they had no recollection of following the incident. Memory is too complex to be explained only by neuron pathways in the brain.

The unit *Citta* is intimately involved with the creation and storage of memories. Hence, memory is not a purely physical phenomenon, and neuroscientists have conclusively demonstrated that there is no specific site for the storage of memories in the brain; rather, memory is psycho-physical. It is dependent on the *Citta* taking on the shape and properties of the original event and a healthy brain to translate these mental vibrations into electrochemical processes of the nerve fibers. Hence, if the brain is damaged, the *Citta* may be unable to express memories in the conscious layer of mind.

The *manomaya kosha* also appears to store an extra-cerebral type of memory—memories that do not depend on the physical brain but are imprinted on the unit *Citta*, occasionally coming to the surface on their own or under hypnosis. The memory of past lives is an example of this type of memory. Children sometimes remember details of a previous physical existence in another body and time. Parents tend to write this off as nonsense and the children forget about their other life after a few years. This forgetfulness is a blessing since memories of a previous life can be confusing and frightening to a child. Hypnosis can bring back memories of past lives. In fact, the standard treatment for neurotic phobias is to hypnotize the patient and regress them to the first time they had the irrational fear.[2] Often they remember an incident that occurred in a previous body and time. Once they relive the traumatic memory, they are normally cured of their neurosis.

The subconscious layer of mind is also responsible for higher thinking, contemplation, reasoning, pleasure, and pain. Philosophies, scientific theories, and all sorts of problem-solving activity take place in the *manomaya kosha*. This layer of mind is active during sleep. Sometimes we go to bed with a problem and wake up having found the solution. Many intellectual and scientific discoveries have occurred when the conscious mind and the sense organs were in a quiet state, allowing a clear awareness of this *kosha*. Insights into problems often occur in the twilight state between wakefulness and sleep.

The next layer of unit mind is the unconscious, superconscious, or causal mind. These three terms are used interchangeably.

The unconscious mind is not bound by time, place, or person. It is the transcendent, all-knowing, timeless, collective mind. For this reason it is sometimes called the superconscious mind and itself consists of three subtle layers—the *atimanasa, vijinanamaya,* and *hiranmaya koshas*. Because the unconscious mind is actually a collective mind shared by all unit entities, in reality it is nothing but the reflection of Cosmic Mind on the individual mental plate or brain. The unconscious mind is responsible for all the higher sentiments and knowledge.

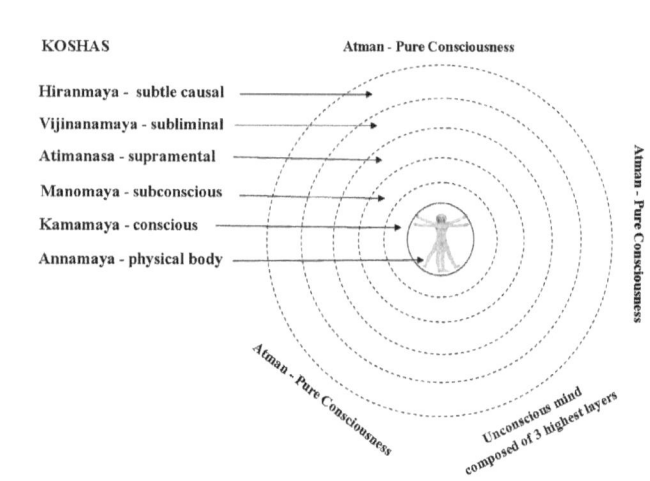

The six layers of mind. The atman is unqualified or pure consciousness and is the witness of all layers of mind.

As noted earlier, the Swiss psychologist, Carl Jung recognized the existence of a higher mind in human beings, which he called the collective unconscious. Mystics and saints tap into the unconscious mind when they experience ecstasy and "awakening." Because this superconscious mind is beyond time and space, past, present and future merge into the eternal now. Prophets tap this resource to make accurate predictions of the future. Accurate knowledge of the past also resides in this mind since the Cosmic Mind-stuff or *Citta* retains an accurate record of every vibration that ever occurred, and the unconscious mind is the portal to Cosmic Mind.

The unconscious mind is the seat for all the higher mental functions. These include magnanimity, humility, serenity, gentleness, mercy, intuition, discrimination, non-attachment or imperturbability, creative insight, sadhana or meditation, and spiritual ecstasy.

As one experiences the deeper layers of the unconscious mind through sadhana, the sense of ego disappears and the "I feeling" diminishes so that one experiences omniscience and a feeling of oneness with the universe. Beyond the unconscious layers of mind lies the atman or unit consciousness, which is beyond duality or any perturbation. As noted above, this atman is identical in quality to the Paramatman, but because it is focused on the unit mind, it is quantitatively inferior.

The mind can be compared to the surface of a pond. Thoughts, desires, emotions, etc. bubble up from below to break up the surface calm. The moon, representing the Paramatman, is reflected as a dancing pattern by waves on the surface of the pond. The atman is witness to the reflection. Wind, representing the breathing and other physical activity, further disturbs the surface of the water. If all this activity were to cease then the mind/water would become calm and reflect the moon perfectly. At this point, the entity witnessing the reflection of the moon (the atman) would see that it was identical to the moon itself (the Paramatman).

8

Happiness, Suffering, Good, and Evil

We have seen how the universe is created from Consciousness, and that the universe is in reality a Singularity or unified Whole. Therefore, every thing, every creature, every human is connected to every other thing, creature, and human. The differentiation we witness is actually an illusion, but knowing this intellectually does not give us the feeling of happiness or free us from suffering.

Cosmic Mind consists of Cosmic *Mahat* (I am), Cosmic *Aham* (I do) and Cosmic *Citta* (collective Mind-stuff). Unit mind results from the reflection of the Cosmic Mind on the unit mental plate. Unit mind is limited, creating the appearance of the individual ego and body. Thus, we are saddled with our individual "I feeling" and ego and become strongly associated and attached to our body and mind. This connection of the unit mind with the body and lower mind is the cause of suffering, and release of the mind from these lower faculties leads to freedom and happiness.

If we ask ourselves what is behind the drive to live life as we do day to day, we would probably say that we are seeking happiness. At the core of our existence is the desire to be happy. If God were to grant us one wish, we would probably ask to experience unlimited and unending happiness. This is natural, for as humans we possess a keen sense of self-awareness or "I feeling." That is, the underlying feeling of "I am" or *Mahat* is the dominant characteristic of our mind. Since *Mahat* is nonlocal and unlimited, our personal "I feeling" is identical to that experienced by every other human. Through this infinite *Mahat,* we share an intimate connection to the Infinite. This connection with the Whole leads to an underlying longing for the Great and to a thirst to experience unlimited happiness.

The longing for limitlessness or the Infinite is the chief distinguishing feature between man and animals. Give a dog a chunk of steak and he will be completely satisfied and soon fall into the bliss of sleep. Treat a human to a six-course meal with wine and he will probably find some fault with the meal. For example, the soup was not hot enough. The truth is, we humans are never 100 percent satisfied with anything on the physical plane because nothing on this plane satisfies our thirst for limitlessness.

The underlying thirst for the Infinite can literally drive people to act in destructive and insane ways when this energy is directed toward the material realm. For example, the sickness of accumulation and greed can occur when a person devotes their life to the acquisition of wealth. After acquiring a million dollars, they remain unsatisfied and unhappy and feel the need to acquire ten million. Once this goal is reached, they still feel unsatisfied and inferior to others who are wealthier. Hence, they devote their energies to acquiring hundreds of millions of dollars or perhaps billions of dollars. After a person has obtained fantastic wealth, they will still feel unhappy and unfulfilled. It seems that nothing less than the acquisition of the entire universe would bring them complete satisfaction. Sometimes, people develop the desire to climb the highest mountain, be the greatest athlete, or excel at some other physical endeavor. Devoting their energy fully on such an extroversive path may ultimately lead to a feeling of accomplishment and happiness upon achieving the goal. However, the feeling is short-lived and never completely satisfying.

The thirst for limitlessness may also be misdirected into the quest for fame or power. The inflated egoism that accompanies this endeavor brings only fleeting happiness; such people never achieve complete satisfaction because no matter how much fame or power they attain it is less than absolute and always short-lived. Some people develop the need to try to become an intellectual giant or the foremost expert in their field. This is not destructive unless it leads to intellectual pride or self-righteousness. Still their energies are focused on the lower plane of existence when they could be directed to the limitlessness of spirit.

On the other hand, there is no limit to the riches of the mind and consciousness. These riches are obtained not through the sense or motor organs, but through introspection—that is, the quest for self-knowledge. Hence, true happiness as opposed to simple pleasure comes when the mind is engaged in the exploration of the limitlessness of the psycho-spiritual realms. True happiness is obtained when the mind is fully engaged in the limitlessness of consciousness or pure being.

In order to experience the unlimited happiness that comes from being absorbed in the Infinite Entity, the mind must first be withdrawn or detached from the external world. That is, the mind must cease being distracted by thoughts and mental images. The process of withdrawing the mind from the external world is the first step in the practice of meditation or sadhana. It is called *pratyahara* in Sanskrit. In the second step, the *Aham* or ego is set aside and the *Mahat* or "I feeling" grows and intensifies. Ultimately the limited sense of "I am this" or "I am that" (ego) is transcended and one experiences only *Mahat*. At this stage, the feeling is "I am God" or "I am one with God." This experience may be called liberation, enlightenment, or self-realization. There is no word in the English language that adequately describes the unlimited happiness or ecstasy that accompanies this state of mindlessness.

Achieving the state of self-realization, even for an instant, is not easy and requires a great deal of dedicated effort and perseverance. The unit is said to move toward the Infinite or Cosmic Consciousness on a path. The path is described as being a difficult path full of pricks and thorns as well as distractions or enticements that can lead one astray. We have seen how the Brahma Chakra impels every living creature toward liberation, but the normal pace of the Brahma Chakra movement is very slow, as evidenced by the timetable of evolution.

Human thirst for the Infinite is insatiable and every human being craves happiness—not a small amount of happiness, but infinite happiness; and not in the distant future but right now. In order to satisfy this thirst for the Infinite, humans need to enter into an accelerated path of self-discovery. As mentioned before, this is the path of *vidya* or knowledge, and it is known as the path of bliss (ananda marga) because taking the path leads to feelings of happiness, and ultimately the path leads to unlimited happiness (in Sanskrit, *ananda*). Chapter 11 has a more detailed description of the practice of meditation or sadhana and how it frees one from the bondage of ego.

As human beings, we possess a preponderance of *Mahat* and are self-aware. We are capable of feeling both great happiness and great suffering. But, what is suffering and why does it exist in this mental concoction of the Supreme Entity or *Paramapurusha*? In other words, if God is good, all-loving, all-knowing, and all-powerful, then why do the units he creates suffer? The answer lies in an understanding of suffering.

God is the ultimate Good, and on the higher levels of mind, we always feel connected to him and bask in the ecstasy of his being. However, on the conscious and subconscious levels of mind the ego is active. We engage ourselves in the mundane tasks of everyday life, and we are normally oblivious to the higher energies that pull us toward the great "I am." The ego is like a parasol that blocks the emanations of the Supreme Entity and prevents our mind from embracing God's grace. There is no reason for the higher levels of mind or *Mahat* to compete with the ego since they are already fully engaged in the Infinite and completely overcome with ananda or bliss.

Therefore, the only mechanism by which the ego can be worn down and our consciousness directed toward the Supreme is by struggle and mental clash. It is inevitable that we encounter these when our attention is directed away from the Supreme. Hence, suffering is a corrective force or teaching tool constantly trying to shift our attention from the crude to the subtle. Suffering can be considered a blessing in disguise. The more we look away from our Higher Self, the stronger the corrective force needed to get us back on track. On the other hand, when we are on the path of *vidya* or knowledge, we feel rewarded with happiness, and pain and suffering have less effect on our mind.

As mentioned in Chapter 6, suffering is governed by samskara or the Law of Karma. Actions that deny our connection to one another and to the One create reactive momenta or samskaras that may result in discomfort, pain, or suffering. In a sense, we are paying ourselves back for our selfish, antisocial behavior. We become cleansed of the negative samskaras when they are experienced or ripened. The experience caused by the ripening of a negative samskara will produce a feeling commensurate with the pain caused by the original act. At an unconscious level, we learn from the negative experience that performing such an action leads not to happiness but to pain and suffering. Hence, we are less apt to perform a similar action in the future.

It is through the ripening of negative samskaras and the experience of clash, pain, and suffering that the ego or *Aham* becomes subtler, yielding more *Mahat*. Great suffering begets great growth. Sometimes people have to undergo great suffering in order for them to change the direction of their lives. A person who devotes all their mental energies to achieving material wealth, power, or fame will necessarily be faced with stronger corrective energies, such as feelings of unhappiness, dissatisfaction, pain, and suffering. This blessing in disguise is the only way the cosmos has of turning such a person away from the false promise of ego attainment to the true happiness of *Mahat* or soul-consciousness. Thus, sages say that the pricks of thorns experienced along the way are a better measure of one's progress on the path of bliss than the sweet smell of flowers.

Almost all unit minds have samskaras remaining in the mind after death, since at the time of death the unit mind will normally have a multitude of unfulfilled desires, attachments, memories, etc. The unit consciousness along with the unit mind must survive the loss of the body since the only way the unit consciousness can be destroyed is by merging and becoming one with the Cosmic Consciousness. In order for the bodiless mind to find expression for these samskaras, it must take on a new physical body. Thus, the Cosmic Mind will find a suitable embryo to serve as the new "home" for the bodiless mind so that it may adequately express its samskaras and move forward on the path to perfection. Hence, the burning of old samskaras and the production of new samskaras may go on for many lifetimes before the individual advances onto the accelerated path of *vidya* in order to complete the journey to perfection.

Because units carry samskaras from one body to the next, we sometimes witness the phenomenon in which a child is born into a life of hardship and suffering due to no obvious fault of its own. Another child may be born with the proverbial silver spoon in its mouth and seemingly "have it made." Since we have no knowledge of a child's past lives, we remain mystified by such so-called accidents of birth. However, everything in the universe is connected. It is no accident that some individuals come into this world with a seeming mountain of problems to overcome while others seem destined for happiness and success. Suffering is never accidental. We bring it upon ourselves in this life or from our behavior in our past lives. The fact that we do not remember our negative deeds from the past does not free us from having to reap their reactions.

Just as there is happiness and suffering in this world, there is also good and evil. In the simplest sense good is the movement from crude to subtle, or *vidya*. *Vidya* means "knowledge" in Sanskrit. Therefore, good action, or *vidyamaya*, is an action performed with knowledge—the action that brings us closer to our goal of becoming one with God. Evil is just the opposite of good. Evil action is termed *avidyamaya*, or the movement towards ignorance or crudity. Such actions create negative samskaras, which lead inexorably to pain and suffering.

The evolution of unit consciousness occurs when the reflection of Cosmic Consciousness becomes clearer and greater in intensity. In this process, the mind becomes subtler and more expanded. This movement toward subtlety (*vidyamaya*) is accelerated by good actions such as selfless service and by focusing the mind on the subtle Higher Self as opposed to the crude material realm. According to the Cycle of Creation, there is an inexorable tendency to move toward the subtle; everyone will eventually have to follow *vidyamaya*, accelerating their movement toward the subtle until they attain the Supreme.

On the other hand, the more the mind is absorbed in crude objects the more unit consciousness is dragged backward, because the reflection of consciousness becomes dimmer with greater expression of bondage. When the mind is absorbed in crudeness, it remains more strongly under the influence of *Prakriti*; as a result, the onward march of unit consciousness is halted. Actions that lead a person to go against the natural flow of Brahma Chakra halt the evolutionary march toward subtlety and result in punishments designed to discourage such behavior in the future. Actions that draw the mind to crude objects, leading one to act against the principle of unity, are by definition evil, while actions that reinforce our connection with the One may be termed good.

The greatest evil would be to deny one's own humanity; that is, to deny one's connection to the One and thus one's connection to everything and everyone in the universe. Such a mindset is no different from that of an animal, which is incapable of contemplating a higher order. But, while animals act primarily under the influence of instinct or Cosmic Mind, man is capable of doing evil deeds that accelerate his movement from subtle to crude. Some truly evil individuals such as Hitler can be considered to have sunk to a state lower than that of an animal. However, even a person whose actions are evil is inexorably drawn toward the limitlessness of the Paramatman. After reaping their negative samskaras, they eventually will leave the path of *avidya* and move forward on the path of *vidya* to eventual unity.

There is a continual conflict going on everywhere between good and evil, light and darkness, virtue and vice. Human society progresses through this conflict. There is no such thing as pure good without an element of evil in it, or pure evil without an element of good in it. Both coexist in every thing and every action. Pure good is only found in the Supreme Consciousness. Since evil is coexistent with good we are always in a state of imperfection. The aim of human life is to progress from imperfection to perfection.

The movement of human society, the movement of both individual and collective life, from imperfection toward perfection, is human progress. Evil is a blind force that puts human beings into the darkness of ignorance. It has to be firmly dealt with by people. The struggle between good and evil went on in the past, is going on in the present, and will go on in the future.

If we accept the reality of the Unity Principle then there is an undeniable purpose behind the creation of human beings. The purpose of the Supreme Entity in creating human beings is to make them follow the path toward the subtle to take them back to the Supreme Entity. This forms the fundamental characteristic or nature of human beings. In Sanskrit, this unique characteristic of human beings is called dharma. To get back to the Supreme Rank, effort is required for the elevation of unit consciousness. The path of *vidya* must be taken, not that of *avidya*, which would lead to more obstacles and less progress. People who work according to their innate nature (dharma) experience happiness and ultimate bliss. Those who take the path toward darkness and work against their dharma do not serve the purpose for which they were created and bring upon themselves unhappiness and suffering.

Some people believe that the results gathered due to evil deeds can be compensated or neutralized by the good samskaras earned by good deeds. This neither happens nor is it possible. It has been discussed earlier that all actions, whether good or evil, cause a deformity in the mind. In the process of mind regaining its normal form the deformity is removed by an equal and opposite reaction. Hence, the samskara or reaction caused by an evil action cannot be removed by a good action. Every samskara is independent of all others and one has to experience the consequences of good and bad actions separately. Thus, the results of good actions cannot help us to evade the suffering caused by bad actions. At best, the mode of experiencing the reaction can be changed: the intensity of suffering can be reduced or increased by slowing or accelerating the speed by which reactions are experienced.

This is similar to taking out a loan for $10,000 to remodel your kitchen. The initial terms of the loan might be to repay the full amount in three years, but you discover that this is impossible to do while putting your daughter through college. The situation may create much struggle and distraction in your mind. This is mostly relieved when the lender agrees to renegotiate your loan allowing you to pay it off in ten years instead of three. The lender may also agree to waive part or all of the payment in return for a service you perform for him, but in no case can you escape payment for the loan. The best way to overcome the consequences of evil deeds is to experience them in the accelerated path of *vidya*. For those on the path toward the Supreme Entity the consequences or samskaras resulting from evil deeds seem to be distributed less harshly lest they distract them from the goal.

The mode of experiencing the reactions to evil actions may be changed with the help of spiritual practice or sadhana. The experience of the reaction or samskara cannot be evaded, but the condition under which the reaction is experienced may be changed. In the example above, a more desirable way of paying the debt might be to perform some service to the party loaning the money in return for a waiver of a part of the amount owed. For example, a plumber might agree to fix a leaking pipe in return for partial payment of the debt. In this way, he does not need to come up with hard cash and may be able to pay his daughter's college tuition for the semester. When either the period or mode for suffering the consequences or reactions of one's actions are changed with the help of spiritual practices, the result may be that one does not feel the same intensity of suffering that might be experienced if one were not on the path of *vidya*.

Just as action done in ignorance or evil begets unfavorable reactions, actions done for the good of others and for society beget favorable reactions. Like deposits in a bank account, all the good reactions or samskaras will have to be withdrawn at some future time. The amount of the experience of happiness and pleasure cannot be changed—only the time required for experiencing it can be increased or decreased. Both the experience of pain resulting from bad actions and pleasure resulting from good actions distract one from the real task of realizing the Supreme. This is why Buddha called these reactions chains of iron and gold. The gold chains might be preferable to the iron chains, but both bind us to this world. Those who carry on spiritual practice or sadhana with the intention of achieving self-realization hope to experience the results of their actions, both pleasure and pain, happiness and agony, quickly so that they may complete the experiencing of reactions in as short a period as possible. When all samskaras are exhausted there is nothing left for the unit to experience except the Cosmic Consciousness. Their liberation is at hand.

The dualistic religions of the West have always struggled with questions about good and evil and suffering. The Old Testament teaches that evil arose in this world when Satan, a fallen angel, spoke through a serpent and seduced Eve into disobeying God's command. The goal of the devil was to lead people away from the love of God and lead them to ignorance or evil. Suffering exists because man has fallen from the grace of God (original sin). The difficulty that Judaism, Christianity, and Islam have with suffering springs from the doctrine that the unit being comes into existence at the time of conception. Therefore, it is an accident of nature if an innocent child is born blind or without arms into a life of great hardship. The unfortunate soul is doomed because man is tainted by "original sin" and the child is a victim of this crime against God.

However, a reasonable person might ask, why would an all-loving, all-powerful God allow suffering and evil in his creation? Moreover, why is there a seemingly random purveyance of God's grace on certain individuals and an apparent curse on others? The simplistic explanation offered by the Old Testament is both unsatisfying and illogical, since if God is all-powerful and the creator of this universe, then nothing can happen that he did not will. If there exists an evil force acting outside of God's will then God would be demoted to nothing more than a demi-god similar to the Greek god Zeus. In many ways, the Old Testament portrays such a god who favors some individuals and groups while taking his anger out on others who disobey his law. Such a god is termed a demiurge. He may fashion the universe but stands outside and separate from it.

Modern Christianity has largely dispensed with this concept of God but still adheres to the doctrine that God's creation is separate from him, as are you and I. However, if there were any limitation to what God can do, it would be to create another God that is separate from him. However, this is exactly what the dualistic religions teach. Man is separate from God and this separation lasts for eternity. The Unity Principle negates the flawed ideology of dualism and offers a logical explanation for why man is drawn to God on the one hand while also being capable of great good and great evil on the other.

9

Life after Death

A grove of aspen trees appears to be composed of individual trees, each one quite separate from the other. However, if one excavates below the surface of the earth one finds that all the trees share a common root system. Science also tells us that all the trees in the grove share identical DNA (genes). Therefore, all the trees in the grove are actually a single organism. Our eyes tell us the trees are separate, but in fact, they are intimately connected with one another; they are truly a single living organism.

The same is true of all living organisms on this planet. On the surface, we appear as separate beings, but underneath we share the same consciousness; we are actually one. This consciousness has no beginning or end. It is neither created nor destroyed. It simply is, and will be forever. We as living beings possess consciousness, and this eternal aspect of our existence confers immortality upon every living thing in the universe. The physical body may fall sick and die, but the unit consciousness that was associated with that physical body can never be extinguished unless and until it merges and becomes one with Cosmic Consciousness.

Thus, life after death is not a "pie in the sky" concept promoted by the religions of the world, but a requirement of the Unity Principle. To reject the concept of life after death is to fall prey to one of the false doctrines of material realism—namely, that there is no connectivity or unifying force in the universe; and all data to the contrary must be fabricated or misinterpreted. Secondly, one would have to reject the testimony of millions of people who have experienced first-hand their connection with the universe or experienced consciousness outside their body.

However, what actually happens in death? The simple explanation is that the mind loses parallelism or touch with the physical body. When an organism is alive, there is parallelism between its unit mind and its body. In higher organisms, the unit mind functions through the nervous system and brain. If the brain is damaged then expression of the unit mind may be impaired or cut off. Physical death occurs when the unit mind can no longer maintain connection with the physical body. This normally occurs during brain death. At this point, the unit mind that is no longer connected with a body is called a bodiless mind.

The bodiless mind contains all the accumulated samskaras and memories of past activities and experiences that were present in the unit mind at the time of death. These mental impressions exist in the unit *Citta* (objective mind-stuff), and are called extra-cerebral because they exist in layers of mind that are not directly dependent on the brain. The bodiless mind cannot feel pleasure or pain. Having no sense or motor organs, it is incapable of sensing or doing anything. It does not actively witness anything and exists in a state similar to deep sleep. The only thing separating a bodiless mind from becoming unified with the Cosmic Mind is the bundle of samskaras associated with it.

In order to express these samskaras the Cosmic Mind will have to find a new body for the unit mind. This could take place almost immediately or it could take many years. Once the Cosmic Mind finds a suitable home for the bodiless mind it will become localized with a fertilized ovum or zygote and eventually mature into a baby and be reborn. The minimum time between death and rebirth for a human being would therefore be nine months. Normally, it takes several years, but the consciousness of the bodiless mind is unexpressed and experiences timelessness like that of deep sleep.

Remembering Past Lives

Extra-cerebral memories include memories of all past events including those that took place in previous bodies. Sometimes a person may recall experiences from a past life. Children usually do not recall much of anything that occurred before the age of three, but sometimes memories of past lives will come through to them. This can cause them anxiety and their parents will usually try to assure them that these memories have no basis in reality. As a child ages, these memories of living in another body fade.

There have been hundreds of cases where parents or others have verified the uncanny accuracy of details provided by children's accounts of past-life experiences such as those of Sita (Chapter 6). Such studies lend evidence to the idea that our consciousness undergoes physical rebirth.

For adults, remembrances of past lives may come out during dreams, deep meditation, or during hypnosis. As mentioned earlier, a common clinical treatment for persons suffering from phobias or neurotic fear is to put them under hypnosis and regress them to a time when they first experienced an incident that caused them to experience that intense fear. Often the traumatic experience they describe under hypnosis occurred in another body and time. Reliving this experience under hypnosis often cures the individual of the phobia.[1,2] Under hypnosis a therapist can guide almost anyone to recall events from a previous life. Interestingly, people recall living normal, unexciting lives, which negates the popular perception that people mostly recall being famous personalities in the past. Except for medical treatment, the remembering of past lives is not recommended since it can create anxiety and uncertainty, and can divert one's attention from the job at hand, which is to know one's self in the here and now.

NEAR-DEATH EXPERIENCES

Another line of evidence for life after death comes from the studies of near-death experiences (NDE). Hundreds of such experiences have been studied and chronicled, many by Raymond A. Moody, Jr.[3] and more recently by Jeffrey Long in his book *Evidence of the Afterlife: The Science of Near-Death Experiences*.[4] One of the most compelling reasons for believing that such experiences offer scientific proof of an afterlife is the remarkable similarity of the accounts, regardless of nationality, religion, race, culture, and other demographics. Dr. Long, a radiation oncologist, asserts that there are nine arguments that prove the existence of life after death. These arguments were generated through the study of consistencies from the hundreds of NDE accounts that he has compiled over the years. These arguments include: (1) how it cannot be medically explained how people experience consciousness outside their body when they are clinically dead; (2) blind people experiencing visual perceptions during their NDE; (3) children giving NDE details similar to adults, though they may have never been exposed to this concept; (4) "life review" experiences that tend to reflect real events. These arguments, along with the others, are the primary basis for Long's assertion that the NDE data prove that there is life after death. One of the most convincing arguments that the mind functions independently of the body during death comes from the accounts of blind people. For example, Larry Dossey tells of one such story of a young woman, Sarah, whose heart unexpectedly went into fibrillation during a routine gall bladder operation.[5] From her vantage point outside her body she could clearly hear and see many details that occurred as the doctors and nurses frantically tried to restart her heart. These included the layout of the operating room; scribbles on the surgery schedule board; color of the sheets; the nurse's hairstyle; and even the fact that her anesthesiologist was wearing mismatched socks that day. Since Sarah was blind from birth, she had never had an experience with sight yet her perceptions turned out to be remarkably accurate.

People who wish to cast doubt on the idea that the NDE shows that our consciousness can function outside our body have put forth the theory that during the process of dying the brain shuts down and we experience a dream-like state as our subconscious mind takes over. They argue that these experiences come naturally during the dying process and are hallucinations.

However, this theory fails to adequately explain how people who are pronounced clinically dead could have such lucid consciousness, a consciousness that is described as more vivid than our normal day-to-day consciousness—nor does it explain many of the other elements of the full-blown NDE. These include: accurate visual and auditory experiences from a vantage point outside the body; the reliving of thousands of past events simultaneously in an altered, omni-view state of consciousness; the feeling of being in the presence of a being who expresses infinite, unending, and unconditional love; and finally the life-changing conviction, even among persons who were previously atheists, that there truly is a God and life after death. The experiences of people that have had a full-blown near-death experience indicate that as the mind leaves the body they come to realize the unconditional love and permanence of their Higher Self and are able to see with the perfect vision of their Higher Self the pettiness, selfishness, ambitions, and illusions of their ego-centered life.

Belief in Reincarnation

The monistic eastern religions universally and wholeheartedly teach the doctrine of reincarnation. In addition, Judaism and Christianity have deep ties to this doctrine. In Judaism, there has been a fundamental belief in reincarnation or gilgul that goes back thousands of years. In modernized versions of the Jewish faith, this belief has been largely ignored. However, it lives on in the Orthodox and Hasidic communities and is central to the Kabbalah or mystical Judaism.

There is ample evidence to suggest that early Christians believed in reincarnation. Gnostic Christians believed that they would be reborn in a new body following death if they were unable to attain perfection in this life.[6] By some accounts, before the rein of the Emperor Constantine the number of gnostic Christians outnumbered the orthodox Christians, but when Constantine decided to make Christianity the official religion of the Roman Empire he favored the orthodox wing because they had an established hierarchy. He also felt that the concept of reincarnation was a threat to his empire since his soldiers might be less inclined to die for the Emperor if they believed they would have to come back again rather than be dispatched directly to heaven. Therefore, most of the early writings and gospels that referred to reincarnation appear to have been deleted from the New Testament in the fourth century during the reign of Constantine.

In the sixth century, the Second Council of Constantinople officially declared that the belief in reincarnation was a heresy. Any teaching of this doctrine was thereafter brutally suppressed. A few references to reincarnation appear to have made it past the censors into the four accepted gospels of orthodox Christianity. For example, Jesus told his disciples that John the Baptist was the reincarnation of the prophet Elijah.[7] And there are several passages in which both Jesus and John the Baptist are thought by some to have previously been Elijah.[8] At that time, the concept of returning to earth in a new body was a commonly held belief, and much of what Jesus taught was that a spiritual change has to take place in us so that we can obtain liberation.[9,10]

PARANORMAL BEINGS

If consciousness survives the death of the body, are stories of ghosts, fairies, angels, and demons real or imagined? Since the bodiless mind does not possess a physical body consisting of any of the five fundamental factors, i.e. ethereal, aerial, luminous, liquid, or solid, it does not possess any nerves or any other faculty that would allow it to affect any physical object. This would not preclude it, however, from affecting a person mentally. However, this too would seem unreasonable since the bodiless mind remains in a nonlocalized, dormant state. The unexpressed samskaras of the bodiless mind that differentiates it from other unit minds and from Cosmic Mind cannot find any outlet for expression until a new physical body is found for that unit mind.

What is behind such stories of the paranormal? Are all such tales hoaxes or imaginary? To answer this question we must first consider the meaning of what is real and what is imagined, or a product of a hallucination. The seat of the sense organs is in the brain and not in the gateway organs. For example, the eyes receive light and transmit nerve impulses to the brain where the actual sensation of sight occurs. This is true of the other four sense organs as well. Thus, everything we perceive actually occurs in our brain and is only indirectly dependent on the external world, which is what stimulates our gateway organs such as the eyes, skin, ears, nose, and tongue. Therefore, nerve impulses that affect our brain but have no link to the external world via our gateway organs can appear real. Such phenomena are often labeled hallucinations and certainly some ghost experiences fall into this category.

Hallucinations are of two kinds: positive and negative. A positive hallucination occurs when thought waves affect the sense organs in the brain and one sees, hears, feels, tastes, or smells something that is not actually present in the external world. One's sense of reality is temporarily impaired, and the conscious mind may get absorbed into the subconscious mind. Unlike the dream state, whose reality we reject the moment we regain normal consciousness, the positive hallucination may seem real even after our conscious mind resumes normal activity. Positive hallucinations may also be elicited by hypnosis. Negative hallucinations on the other hand occur when the mind refuses to see something that is actually present. These can also be brought on by outer-suggestion (hypnosis) and by autosuggestion, most commonly when, because of fear, one's mind refuses to accept some aspect of an experience.

Fear has a powerful effect on the mind. It can cause temporary concentration of mind. For example, if a person believes in the existence of ghosts and visits a house that is said to be haunted, fear may trigger a positive hallucination of a ghost. The concentration of mind caused by fear can leave an imprint in the *Citta* of the Cosmic Mind at that location. That in turn may trigger the subconscious minds of other people and cause them to experience similar hallucinations. Therefore, most ghost stories are propagated by fear and have no physical reality.

However, if the psyche of a person is disturbed enough it potentially can move physical bodies. Stories of entities moving physical objects may be due to the action of *Citta* on physical matter. However, such activity is not the result of the action of a bodiless mind and is wholly dependent on the mental action of a living human being.

There is another category of paranormal sightings that are explained by what in Sanskrit are called *devayonis*. These are luminous bodies, or advanced souls who were unable to attain liberation due to some strong desire or attachment that remained in their mind at the time of death. Suppose, for example, that a person meditated regularly for most of their life and their mind became very subtle and spiritually advanced. In addition, suppose that during their life this person had a great love of the fine arts and in particular for the intoxicating rhythms of fine music. At the time of death, this person may lack the necessary devotion to be able to surrender their unit consciousness completely to the Supreme Consciousness due to their deep desire to continue to enjoy music. The mind of such a soul or unit consciousness may be too subtle to be reborn directly in a normal physical body consisting of the five fundamental factors. Instead, they take on a body having only ethereal, aerial, and luminous factors. Lacking a complete physical body with liquid and solid factors, they will be unable to perform sadhana and continue in a normal manner on the path to liberation. They will be able to enjoy the vibration of music sympathetically by frequenting concert halls and other places where music is played. However, until their taste (samskaras) for music is satiated, which could take many years, they will remain as a luminous body and will not be able to be reborn in a physical body.

Sages have identified seven different types of *devayonis*, according to their different desires. The most highly developed *devayoni* is the siddha. A siddha is a spiritually advanced soul who had a desire at the time of death to be reborn with spiritual powers so as to do great work. Such beings enjoy sympathetically the vibrations emanated by persons doing meditation and by persons who are highly advanced spiritually. Since *devayonis* have the luminous factor, they may at times be perceived with the naked eye. Their presence may also be felt under the right circumstances. Many sightings of angels, fairies, and friendly ghosts may in fact be *devayonis*. It is possible that some sightings of UFOs are luminous bodies.

10

Psychic Body and Power

True power lays in the subtle not the crude. A weight lifter may be able to press two hundred kilograms, but if a tiny nerve is cut, he may be unable to lift a cup of water to his mouth. Hence, muscles are dependent on nerves, which in turn are dependent on the brain. And, what is the brain dependent on? The mind or psyche.

Matter and energy can be considered complementary aspects of the same thing referred to in Einstein's equation: energy equals mass times the speed of light squared. In other words, matter is simply a condensed form of energy, and today scientists refer to these as mass-energy. Even a very small change in mass is accompanied by a very large release of energy. Consider for example the energy released when a few grams of heavy hydrogen nuclei are forced together to form helium in a thermonuclear or hydrogen bomb. However, the energy locked up in all the mass in the universe pales in comparison to the energy of the void. Scientists call this "dark energy" and have calculated that it accounts for roughly three-fourths of the total mass-energy of the universe.

The void or empty space is nothing but the ethereal factor—the subtlest of the fundamental factors that make up the universe. At present scientists are puzzled by dark energy and are unable to explain what it is or how it arose during the Big Bang. Scientists also know that the universe contains much more mass than is observable. They call this dark matter.[1] Ordinary matter-energy accounts for only 5 percent of the known universe and the two mysterious "dark" substances apparently comprise the remaining 95 percent of the universe.

The Unity Principle dispels most of this mystery. It postulates that mind is a condensed or cruder form of Consciousness and that the fundamental factors are formed from Consciousness by additional binding. Everything in the universe originates from Consciousness; it is the source of all power, energy, and matter. In the hierarchy of Consciousness followed by mind and then matter, there is diminished power and energy associated with the cruder factor. This is also true of the fundamental factors and explains why there is more energy in the void of space than in all the matter of the known universe.

Every living organism, even the most primitive, displays traces of mind and consciousness. For example, a paramecium (a single-celled protozoan) can be observed in a microscope to swim about swiftly searching for food. If it bumps into an object, it recoils and darts off in another direction. Similarly, euglenas (unicellular protist) have an eyespot, a primitive organelle that is sensitive to light, and they are able to adjust their position in order to produce more food via photosynthesis. More evolved multicellular organisms show greater and more complex mental capabilities.

The unifying force that allows these organisms to function and survive is their unit mind and associated psychic body. This psychic body is their connection to the Cosmic Mind, and it is through this connection that they are guided instinctively. Without a psychic body it is hard to explain how a diverse collection of billions of individual cells can work together to form a viable organism such as a bird or a dog.

The Chakras

The psychic body of a living organism is closely related to its physical body. For every physical organ, there must necessarily be a corresponding psychic one. In vertebrates, the most important controlling centers or plexuses of the psychic body lie along the spinal cord and in the brain. They are called chakras in Sanskrit, which literally means "circle." There are seven chakras in the human body. These are: (1) the *muladhara* chakra at the base of the spine, above the perineum; (2) the *svadhisthana* chakra, controlling the genital organs; (3) the *manipura* chakra at the navel or solar plexus; (4) the *anahata* chakra, associated with the heart; (5) the *vishuddha* chakra at the throat; (6) the *ajina* chakra, located between the eyebrows and controlling the mind; and (7) the *sahasrara* chakra at the crown of the head—it controls all the other psychic centers.

The first five chakras are the controlling points of the five fundamental factors within the human body. The *muladhara* controls the solid factor, *svadhisthana* the liquid, *manipura* the luminous, *anahata* the aerial, and *vishuddha* the ethereal. The unit mind is a minuscule clone of the Cosmic Mind and is controlled by the *ajina* chakra. The highest chakra, the *sahasrara*, is called the thousand-pedaled lotus because it controls the other chakras and the one thousand propensities of the physio-psychic body. This chakra is also the seat of pure unit consciousness and one's connection with Supreme Consciousness.

The chakras are each associated with corresponding physical nerve centers or plexuses as well as with major glands of the endocrine system. They are also associated with sub-glands like the liver, spleen, pancreas, etc. They control the flow of psychic energy in the body. The non-physical, psychic nerves of the body are termed *nadis* in Sanskrit and the psychic or vital energy they transmit is called prana.

Each of the chakras has a color, certain sounds, and a number of flower-like petals associated with it. These fundamental sounds number fifty in all, denoted by the fifty vowels and consonants of the Sanskrit language. Each petal is believed to control one of the fifty major human propensities, such as anger, hate, desire, love, etc. The five *koshas*, or layers of the mind—*kamamaya, manomaya, atimanasa, vijinanamaya* and *hiranmaya*—chiefly control the five lower chakras, the *muladhara, svadhisthana, manipura, anahata* and *vishuddha* respectively. The greater the control one achieves over one's *koshas* through spiritual practice, the more control one obtains over one's organs. Thus yogis who spend many hours performing asanas (yoga poses) and meditation may be able to gain extraordinary control over their bodies and minds as they become established in the higher *koshas*.

The base or *muladhara* chakra controls the crudest factor—solid. Solid represents the lowest point in the Cycle of Creation. It is the point where *Prakriti* is at her strongest and at which the potential for spiritual growth and expression is greatest. The *muladhara* chakra is thought to be home of the kundalini, the dormant or sleeping spiritual potentiality. It is believed that spiritual practices can awaken this serpentine-like energy, and as it passes upward through the principal psychic nerve or *nadi* that corresponds to the spinal cord, it pierces the various chakras and progressively leads to greater and greater divine experiences. When the kundalini reaches the uppermost chakra, the *sahasrara*, the unit being unites with the Cosmic Entity.

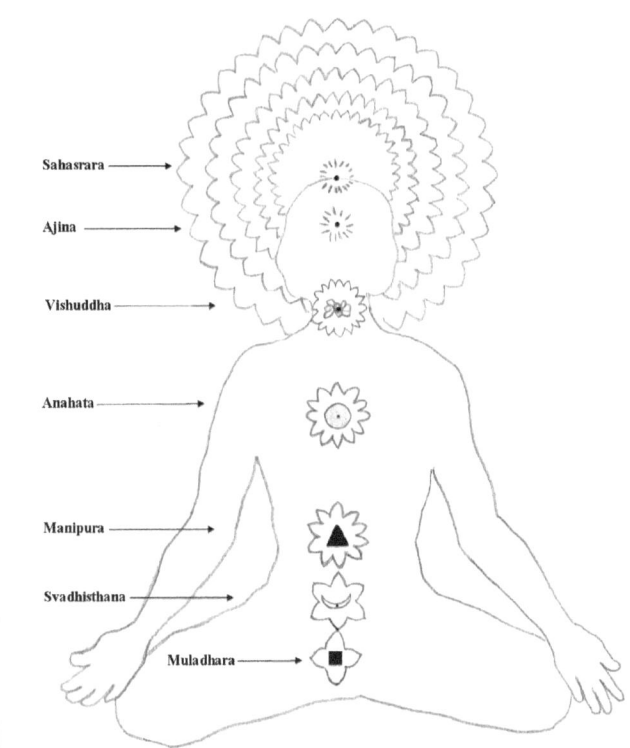

The seven principal chakras of the human body.

It is apparent from the above discussion that the psychic body has tremendous spiritual significance. Only human beings possess the extremely complex psychophysical structure that enables them to be fully self-aware and perform the spiritual practices necessary for moving on the *vidya* path to perfection. Animals have psychic bodies but their physical brains and corresponding psychic structures are more primitive and do not support a high degree of self-awareness. Therefore, animals are unable to perform sadhana and move on an accelerated path toward liberation.

It is with the help of the crude physical nerves and corresponding psychic body that the *koshas* connect and control the different chakras as well as the propensities belonging to them. Without the nerve fibers, it is not possible for the mind either to connect to the physical body or to function physically. During death both the psychic body and the physical body dissolve. The bodiless mind of a dead person, in the absence of the psychic body and physical nerve cells, is unable to express thoughts or emotions, or to affect anything physically. Thus, the bodiless mind cannot contemplate the Cosmic Entity, express its hopes and desires, or become involved with any living entity or crude object. It must become associated with a new physical body in order to progress on the *vidya* path to perfection.

Psychic Powers and ESP

Parapsychologists have identified several extrasensory or psychic powers, including clairvoyance, telepathy, precognition, and telekinesis. These are also known as ESP (extrasensory perception). Clairvoyance or remote viewing is the ability to obtain information about places, things, or events at a remote location. Telepathy is the transfer of thoughts or feelings between individuals at a distance without any apparent physical means. Precognition involves perceiving information about future events before they occur, and psychokinesis is the ability of the mind to influence matter, space-time, or energy.

There may not be any real difference between clairvoyance and telepathy since a clairvoyant may be getting the extrasensory information from another individual. Similarly, it is difficult to prove that precognition is actually different from psychokinesis. For example, if a person is able to predict the output of a random number generator it is also possible that their mind is influencing the generator.

Thousands of well-controlled scientific studies have provided incontrovertible evidence that humans possess ESP. The best evidence comes from statistical analyses of large groups of ordinary people and from meta-analyses of a large number of studies of similar design. The size of the effect is normally small but is consistent across time and experimental designs resulting in a statistically significant effect that cannot be explained by chance, fraud, or elimination of non-conforming data. As mentioned previously, the best review of the data validating the existence of ESP is probably in the recent book by Dean Radin (*The Conscious Universe: The Scientific Truth of Psychic Phenomena*).[2] Another line of evidence validating the effect of mind on matter comes from studies of Intention Imprinted Electrical Devices (IIED). In this experiment a specific intention is imprinted into a simple electronic device to which is imparted (by an experienced meditator) the intent to affect a measurable change in some experimental condition. These devices influence a particular target experiment simply by putting the device close to the experimental apparatus and turning it on.[3]

Generally, ordinary people possess no more than a tiny capability for ESP and it takes a large number of people and statistical analysis of the data to demonstrate an effect that cannot be explained by chance alone. Occasionally a person will come along that possesses extraordinary psychic abilities—a so-called psychic savant. Edgar Cayce was one such individual. Such individuals are rare, but their ESP abilities defy all explanations purveyed by skeptics, atheists, and materialists. More about the life and extraordinary abilities of Edgar Cayce were discussed in Chapter 3.

All psychic powers depend on the fact that the subtler layers of mind are nonlocal. By tapping into the unconscious or superconscious layers of mind we can know and experience things beyond our sense organs and thus beyond our individual brain activity. Telekinesis, on the other hand, involves moving physical objects with the mind. The mystery of how the mind can influence a physical object is no different from the unknown way by which our mind moves our hand. Mind is non-physical and in the absence of some mechanism that allows it to connect or influence matter it would be unable to affect anything in the physical realm.

The most reasonable explanation of how mind and matter interact is that they connect at the quantum level. For example, it is known that a quantum particle can simultaneously exist and not exist; exist in several different states at the same time; or be spread throughout the entire universe. This is required by the uncertainty principle. However, when the particle is measured or observed by any means, its state is no longer uncertain and this act of measurement instantly forces it into just one state. This is termed the "collapse of the wave function." Experimental studies have proven repeatedly that consciousness or mind affects the fate of quantum particles by fixing them into one particular state. Apparently, mind, which is by definition non-material, affects matter in this way. When the quantum wave function collapses by mind or consciousness there is a fixing of matter in momentum, energy, position, or space-time. This is the logical mechanism by which mind or consciousness influences matter. In the final analysis, this interaction is made possible by the fact that matter is formed from mind and not vice versa.

Quantum mechanics tells us that all quantum events have uncertainty and this uncertainty is dispelled by observation. The universe is created from Cosmic Consciousness and it is transformed into Cosmic Mind, which permeates but transcends space-time. Therefore, it may be argued that there really is no uncertainty at the quantum level since mind is always present to fill the uncertainty gap. Therefore, Einstein was probably correct when he argued that God does not throw dice. Everything happens for a purpose; nothing is by chance since this entire creation is nothing but the internal mental concoction of the Supreme Entity. Events may appear random or coincidental to the casual observer but only because we are unable to see the big picture. Experimentally, uncertainty on the quantum level is a fact of nature, but this uncertainty only occurs when we ignore an important complementary aspect of quanta—mind.

Occult Powers

Spiritual aspirants on the *vidya* path of knowledge may develop certain psychic or occult powers because of their practices. These may include various abilities such as levitation, psychic travel, etc. These powers or siddhis are not the goal of spiritual practice and may actually distract one from the goal, which is to merge with Supreme Consciousness. There is a story of a yogi who had developed the power of levitation. He demonstrated his ability to a large group of people by walking on water across the Ganges River. Most of the observers were greatly impressed with this show of occult power, but one observer was not impressed. He turned to his friend and said that he thought the value of this power equaled only two rupees—the cost to take the ferry across the river.

One problem with displaying or using such powers is that they may inflate the ego of the practitioner. People may be impressed by a person with such power and begin to look up to and even worship that person as a great teacher or guru. Unless that person is fully realized like the Buddha, such admiration may lead to an inflated self-image, which is exactly the opposite of what is needed for success on the *vidya* path. Another problem with using occult powers is that they inevitably make the mind cruder and are eventually lost if used continuously. The power to heal the sick may seem beneficial; however, it can involve taking another person's samskara onto oneself, which can potentially rob that sick person of a needed life lesson as well as adding samskaric burden to the healer.

Some individuals perform spiritual practices for the sole purpose of acquiring occult powers. Others may stray from the path of knowledge and devote their energies to expanding their ego rather than powdering it down. Such individuals are termed *avidya* Tantrics in Sanskrit (the Sanskrit preface *a* signifies negation or opposition). They can potentially do great harm to society. For example, Hitler may have been an *avidya* Tantric. He was fascinated by the occult and apparently had the power to influence people almost hypnotically. For this reason, spiritual teachers or gurus are very careful whom they take on as students and require that their disciples be well ensconced in morality, have a yearning and love for the Supreme Entity.

11

Meditation

Meditation is known as sadhana in Sanskrit. It is from the root word "*sadh*," which means, "to try." Sadhana is spiritual practice or intuitional practice. A sadhaka is one who is engaged in the practice of sadhana—a seeker. In the broadest sense, meditation is the process in which the unit "I feeling" is associated with the Cosmic "I feeling" or *Mahattattva*. In this process, the ego is set aside. Thus, when a musician is "lost" in playing his music or a scientist is fully engrossed trying to solve a problem, they are meditating. Their sense of doership disappears as their mind becomes detached from their body.

In a spiritual context, meditation may be defined as any practice in which the mind is directed away from the crude world of the sense and motor organs toward the Supreme Subjectivity or Cosmic Entity. In true meditation, the mind is concentrated or one-pointed. It is not the same as deep or concentrated thinking since in these there is no attempt to associate the unit "I am" with the Cosmic "I am."

Meditation or sadhana broadens and expands the mind. How? It is the innate characteristic of the human mind to become as it thinks. Since there is no greater entity than Supreme Consciousness, when the mind ideates on him, it expands. If the mind focuses on a crude object like an idol or money then it tends to take on the crude qualities of the physical object and it contracts.

Sadhana is the key to *vidya* or the accelerated path of knowledge. Devoid of spiritual practice, the individual can only move slowly on the path toward perfection—learning from past mistakes and expanding their mind slowly by the constant clash, and resulting cohesion that comes from life experiences. In contrast, by practicing sadhana the individual can experience their samskaras rapidly as their mind expands with greater *Mahattattva* or "I feeling" at the expense of the ego or *Ahamtattva*. In the end, the Unity Principle requires that there be only one "I feeling" in the cosmos. By tuning into one's own "I feeling" one is actually drawing closer to the Supreme Consciousness. Ultimately, the little "I" will disappear during sadhana leaving one with the feeling "I am God." This state of pure being is called samadhi in Sanskrit. In English, we sometimes call it liberation, self-realization, enlightenment, or perfection.

Tantra

Sadhana is central to Tantra. Tantra literally means "the practice that overcomes crudity and bondage." Anandamurti reports that some Tantric practices began before the time of Sadashiva, who lived approximately seven thousand years ago.[1] However, Sadashiva systematized and refined the spiritual practices of Tantra and introduced Tantric sadhana to a few spiritually mature followers called *avadhutas*. Over the ages Tantric practices have been refined and improved through experimentation and intuitional knowledge. The three main schools of Tantra are Jain, Buddhist, and Post-Shiva (*Shivottara*). The principal differences among them are in their terminology.

Morality and Dharma

In order to perform spiritual meditation or sadhana a person must observe moral precepts. Without morality, sadhana is impossible since the mind will be dragged down by the weight of immoral actions, which create reactions or samskaras that prevent it from concentrating. Morality is therefore necessary on the *vidya* path, and most of the world's religions put great emphasis on thinking and acting morally.

Acting morally is thus the first step on the path, and sadhana is the second. Together they comprise what in Tantra is called human dharma. Human dharma is the one defining characteristic that makes us human—our innate desire for perfection. By performing spiritual practices, we can reinforce our humanity and truly know ourselves.

The purpose of the creation, the purpose of human beings, is to know who and what we are.

MEDITATION PRACTICES

Paramapurusha is the Supreme Subjectivity; that is, he sees everything. Everything is within his mind. He is the Supreme Subject and everything else is his object. The Cosmic Consciousness is Seer (the subject) and the seen is the object. Hence, this Entity cannot be the object of meditation. Instead the meditator thinks, "I am the object and he is observing me." In fact, there is a fundamental truth to the idea that God is omniscient, and he observes everything we think and do. The Unity Principle requires that the Supreme Entity must observe whatever we observe in our mind. The unit consciousness or atman observes every thought, dream, sensation, feeling, etc. that we experience. He is the witness of everything we experience, and since the atman is nothing but that part of the Cosmic Consciousness that is reflected onto our mental plate, he, the Paramatman, is the ultimate witness.

The constant reminder that we are but a small wavelet in the mind of the Supreme and that all other persons are nothing less than manifestations of that same Infinite Entity can be a form of sadhana in itself. Normally, however, sadhana is performed while sitting comfortably, and the practice may take many and various forms. One of the simplest forms of meditation is to concentrate on a particular object like a candle flame.

A more effective practice is "mindfulness" meditation. In this practice, the focus of the mind shifts from attention at its surface, which is in a constant dance of change, to a deeper experience of silence and calmness. This state of suspension is the domain of the "I am" or the source of everything. Hence, mindfulness means being aware of the observer in us. To perform simple mindfulness meditation, choose a place that is free from external distractions. Sit comfortably in a chair or couch, preferably cross-legged. Close the eyes and let the mind focus on the breath. Try to remain in the here and now and let any thought, image, or sensation come and go freely without trying to push it out of your mind or paying any attention to it. As distractions come into your mind, bring your focus back to the anchor of your breathing. Try to feel that when you breathe in you are taking in the blissful void, which permeates the cosmos, and when you breathe out you are returning the totality of your being to the Infinite.

One of the most effective methods of meditation is using a mantra. Mantra is a Sanskrit word; it means "that quality that liberates the mind." Hence, mantras are typically Sanskrit words or phrases that have particularly strong spiritual vibrations and a meaning associated with them (often a name of God). A mantra may be repeated silently (*japa*) or sung out loud (kirtan). The mind is naturally attracted to the sweet vibration of the mantra, and ideally, it is drawn away from the chaos of the external world into a deep state of concentration and peace. A mantra may be considered a spiritual weapon—able to cut through the baggage of the ego and conscious mind and open a door to the subtle layers of mind and the great "I am" within.

The two most common types of mantra used for meditation are *biija* mantra and *ishta* mantra. A *biija* or seed mantra may be one of the sounds associated with a chakra. By repeating the mantra silently, the mind is drawn to a deep layer below where thoughts originate. The mantra is also designed to open up a chakra, such as the heart center, allowing love to flow through ones being.

An *ishta* mantra is central to the practice of Tantric sadhana. It is a two-syllable Sanskrit word meaning "God." The first syllable is repeated silently while breathing in and the second is repeated while breathing out. It is necessary to go to a teacher of meditation to receive an *ishta* mantra since it is customized to each person's mental and spiritual propensities. An *ishta* mantra is imparted with strong spiritual vibrations by a spiritual master or guru (Sanskrit: *gu* + *ru* = dispeller of darkness). This mantra is especially designed to cut through the outer layers of the mind and hit or excite the dormant spiritual energy or kundalini that lies dormant in the lowest chakra. This energy then rises up the principal channel of the psychic body, purifying each of the chakras above it, and ultimately bringing spiritual energy and bliss to the practitioner.

Before sitting for meditation a spiritual aspirant can often get in a meditative mood by singing kirtan or practicing yoga postures. Another preparation for repeating the *ishta* mantra internally is for the spiritual aspirant to withdraw the mind from the sensory and motor organs, the body, and the physical world. This process is termed *pratyahara* in Sanskrit. *Pratyahara* is an important component of Tantric sadhana because the mind needs to be detached from the physical universe before it can become fully engrossed in the very subtle vibration of the *ishta* mantra. However, once the propensities of the mind are withdrawn from the objective world, the mind is directed to an *ishta* chakra selected for the aspirant by the guru. When practiced properly, this type of sadhana results in stimulating and raising the dormant spiritual energy or kundalini and transforming unit *Citta* into *Ahamtattva* and *Ahamtattva* into *Mahattattva*.

Another Tantric meditation practice involves visualization. This technique is termed dhyana. Tantric yogis and Tantric Buddhist monks—especially in Tibet—practice dhyana meditation. The form selected for ideation is normally that of the guru. Although the Supreme Entity cannot be objectified, the guru is considered by spiritual aspirants to be the crystallized form of the Higher Self. The guru is a door to the Cosmic Entity. The spiritual aspirant or sadhaka finds it almost impossible to visualize the subtle form of the guru. In the process of trying to visualize the form, the mind expands and becomes more in tune with the Higher Self. The practice of dhyana helps the sadhaka develop deep love for God, which is a powerful force for powdering down the ego.

The unit mind is naturally attracted to the Supreme Consciousness. This attraction is analogous to the gravitational attraction of a planet for its sun. If the motion of the planet is slowed, it will fall closer to its sun. Similarly, the meditator wishes to slow their mind and let it fall into a closer orbit around the Cosmic Nucleus. Eventually this will result in a complete loss of action—the unit mind will merge with the Cosmic Entity and experience the condition of ultimate peace or samadhi. This truly is the sole purpose of human existence. Meditation when performed properly is the principal tool to achieve this goal.

Waves on the ocean of consciousness.

12

The Unity Principle in the Teachings of Religious Prophets

The concept that the Transcendental Entity manifests as this world is probably older than human history, and reference to this truth can be found in almost all the world's religions and religious writings. It has been called "The Perennial Philosophy," a term that was popularized by Aldous Huxley in a book by the same name. The Unity Principle probably had its earliest origins some five thousand years before Christ.

Sadashiva (~5000 BCE)

Human beings began to evolve on this planet about one million years ago, but human civilization is at most fifteen thousand years old. Religion had its beginnings about this same time, as described in the earliest religious text known—the Rig Veda. By seven thousand years ago, when Sadashiva (Shiva) was born in what is today India, the human society had not yet invented script and was still undeveloped. People led a simple life. It was, however, a turbulent, conflict-ridden time in India. The indigenous people clashed with the Aryans, who were outsiders that were beginning their migration into the subcontinent.

Anandamurti reports that Shiva made significant contributions in the areas of music, dance, hand and body gestures (mudras), system of marriage, science of breath control, medicine (Ayurveda), and ethics.[1]

However, Shiva's greatest contribution to humankind was in the arena of self-knowledge—the path of attaining salvation. The Tantra of Shiva was a collection of his wife Parvati's questions (*nigama*) and Shiva's answers (*agama*). He taught that self-knowledge and hence liberation was attained by the practice of sadhana. He stated that this universe originated from a conscious Entity, which is maintained in the vast body of that Cosmic Entity, and that every living creature will eventually have to merge again in that Supreme Entity. Hence, Shiva was the earliest proponent of the Unity Principle.

Shiva apparently possessed occult powers from his childhood. Some persons attain divinity by performing intense and prolonged sadhana, for example, Buddha. However, Shiva was an embodiment of Divinity from the time of his birth. He had no spiritual teacher as such. He was the original preceptor or guru, and therefore he may be considered the father of spirituality. The monotheistic concept of one God and the panentheist vision that all things spring from and are manifestations of the Transcendental Entity seem to have originated with him.

In Tantra the entire creation is known as *sambhuti*, and when Brahma with the help of the five fundamental factors takes physical form it is called his *mahasambhuti*. Shiva appears to have been such a personality—an incarnation of the Supreme Entity. Shiva had all the qualities of a *mahasambhuti*—unlimited flow of intellect, unprecedented wisdom and knowledge, unconditional love for all living things—and he was revered as a leader and father figure. Not surprisingly, following his death Shiva was accepted as a divine personality and was worshiped as a god. Even today, there are many followers of Shiva known as Shaivites, and he is considered by them to be God incarnate.

Lord Krishna (~1500 BCE)

The Sanskrit verb *krsh* means, "to attract or draw everything to one's self." Combined with *na* it gives us Krishna, which means "the being that attracts everything of the universe toward his own self." Krishna was born in India some 3500 years ago when the country consisted of a number of diverse kingdoms that lacked unity despite their social and cultural similarities. Lord Krishna's mission in life seems to have been to unite the warring factions and create a Great India or Mahabharata. According to the epic story the Mahabharata, he accomplished this through a campaign of war and unification. The well-known Bhagavad Gita, or Divine Song, is from the Mahabharata.

In the Bhagavad Gita, Krishna reveals himself as God to his disciple Arjuna when he says, "I pervade and support the entire universe by a very small fraction of my divine power." He also explains that he incarnates himself in this world from age to age for the protection of the virtuous, the destruction of the wicked, and the restoration of dharma. This is the role of a *mahasambhuti*, and Krishna as Shiva appears to have been born fully realized from the time of birth.

Krishna taught that this world is a creation of maya, the powerful force of illusion that creates confusion and distinctions. The way to overcome this powerful force of distraction, he says, is to surrender completely to the Lord. In this way, the ego may be transcended and the individual can attain God-realization. Following his death Krishna was revered as one of the personified forms of God in the Hindu religion.

The cosmic vibration of a great personality or *mahasambhuti* may last for millennia, and even today, Krishna is the object of devotion for millions of people. His philosophy and teachings were monistic and if he were a *mahasambhuti* then he would be nothing less than the embodiment of the Unity Principle. The Tantric concept of *mahasambhuti* is very similar to that of avatar. In Hindu scripture, an avatar is considered God incarnate.

Moses (~1400 BCE)

Moses was one of the principal prophets of Judaism. His main contributions were in the arena of morality (Ten Commandments) and monotheism. The teachings of Moses in the Old Testament seem to bear the mark of duality rather than monism. Although it is not entirely clear whether Moses taught that God was omnipresent as well as omniscient and omnipotent, Judaism certainly evolved toward this understanding of God, and such a description of the One God present everywhere and in everything is consistent with the Unity Principle.

Confucius (~550 BCE)

Confucius was a great teacher and philosopher that lived in what is now modern-day China in sixth and fifth century BCE. Confucius taught the philosophy of humanism—that man could cultivate virtue and develop moral perfection. Confucius taught morality and a positive philosophy of life but apparently had little to say about spirituality. Hence, Confucianism is usually not considered a religion or spiritual path, yet its followers were greatly influenced by Chinese folk religions and Taoism. It seems that Confucius never addressed the concept of the Unity Principle.

Lao Tzu (~500 BCE)

Lao Tzu is usually regarded as the founder of Taoism and is credited with writing the Tao Te Ching, the fundamental scripture of Taoism. Taoism and Confucianism are the two most important social-spiritual traditions originating in China. Tao means "the path or way of life." It can also mean the starting and end point of creation—the *Purushottama*. Taoism teaches that to know God, one must know oneself and that this may take many life times. It describes all change as the interplay between the two complementary forces of yin and yang. Although these forces are considered opposite, they are complementary parts of the One. The yin-yang symbol (see below) represents these two complementary opposites within the Whole. Although yin and yang describe the duality inherent in nature they spring from the Whole (Tao). Thus, Taoism affirms the Unity Principle. Duality is an illusion. Truth lies in the One. Knowledge of the One is the only true knowledge.

THE UNITY PRINCIPLE IN THE TEACHINGS OF RELIGIOUS PROPHETS 123

The yin-yang symbol.

GAUTAMA BUDDHA (~400 BCE)

The Buddha was born as Prince Siddhartha Gautama in what is now modern day Nepal at a time when the teachings of religion or dharma had deteriorated to lifeless, ritualistic practices. The story of Siddhartha's life begins with an astrologer telling his father, the king, that his son will become either a world ruler or a great spiritual master. This led his father to try to insulate his son from knowledge of the existence of sorrow and death—lest he desire to take a spiritual path. Therefore, Siddhartha grew up as a pleasure-loving youth that eagerly enjoyed all the sensory pleasures that life could offer.

One day Siddhartha ventured from his luxurious palace and happened to see a feeble old man hobbling with a cane. Having never witnessed any evidence that people aged, became sick, and died, he demanded an explanation from his associates and learned that eventually everyone succumbs to old age and dies. Next he met a monk who told him that the way to attain that which does not change with time is to renounce one's worldly possessions and the short-lived world of sensory pleasures and devote one's entire being to the attainment of God. Shortly thereafter Siddhartha renounced his kingdom, family, and worldly possessions, and began his long and difficult journey to become the Buddha, the Enlightened One.

After attaining enlightenment, the Buddha acquired a large following of people that were inspired by his message of morality, compassion, and the Eight-Fold Path to enlightenment. He opposed the ritualistic religions that were popular at the time and denounced asceticism as unnecessary and a wrong way to approach the Ultimate Truth.

Buddha did not explicitly mention the Supreme Consciousness or God, but it is also true that he never denied him. Many believe that Buddha refrained from answering questions about the existence of God because he knew that God's existence is beyond the scope of the human mind and therefore should not be discussed. Instead of preaching about Supreme Consciousness, Buddha taught *Shunyavada*, the Doctrine of the Void. According to him, everything of this universe emanates from *Shunya*, everything is maintained in *Shunya*, and finally everything will merge in *Shunya*.

Buddha strongly supported the doctrine of rebirth. He taught that our journey in this world is an endless succession of birth, life, death, and rebirth until we ultimately attain liberation (*mahanirvana*). Buddha also said little about the unit consciousness or atman, however, since he taught that man must be reborn in human form, it follows that man must possess a unit consciousness, for without it rebirth would be impossible.

Buddha taught four principles called the Four Noble Truths. First, there is suffering, second there is a cause of suffering, third there is cessation of suffering, and finally there is a way to permanently end suffering. By emphasizing suffering as central to the human condition, he was speaking to people of that time who were accustomed to hardships and suffering in their lives. He could just as easily have formulated these truths around happiness since both are conditions of the mind, and it is only when the mind is suspended that both conditions disappear and one experiences ananda, indescribable bliss.

Buddha taught the Unity Principle to his followers. His Eight-Fold Path of human progress was designed to lead humans from crudity to the Supreme Subtlety. The steps on his *vidya* path consisted of (1) proper philosophy, (2) proper determination, (3) proper speech, (4) proper occupation, (5) proper exercise, (6) proper work, (7) proper meditation, and (8) proper attainment of samadhi or union.

Patanjali (~150 BCE)

Patanjali was a great philosopher and yogi. He compiled the Yoga Sutras, which delineated Ashtanga Yoga or the eight-limbed path of yoga. These eight limbs were *yama* and *niyama* (do's and don'ts, or ethics), asanas (postures of hatha yoga), pranayama (breath control), *pratyahara* (withdrawal of mind), *dharana* (concentration), dhyana (meditation), and samadhi (suspension of mind or absorption in the One). Patanjali said the human mind has fifty main propensities, and these propensities work in ten different directions. Each propensity works within and without, internally and externally, thus fifty times two, or a hundred ways. In addition, these propensities work in all ten of the sense and motor organs, or one thousand total ways. He said that these one thousand propensities are controlled by the *sahasrara* chakra or thousand-pedaled lotus associated with the pineal gland.

According to Patanjali, one obtains samadhi when all one thousand propensities of the mind are controlled or suspended. In this state of suspension, the unit consciousness unites with Supreme Consciousness. This is yoga. The analogy is that of a river running into the sea. It no longer remains a river—it becomes one with the sea.

Jesus of Nazareth (~4 BCE–29 AD)

Most of what we know about the life and teachings of Jesus are from the four canonical Gospels and the gnostic Gospels, most of which were unknown until 1945, when the Nag Hammadi texts were unearthed near Alexandria, Egypt. Most historians would agree that Jesus was a Jew who was born and raised in Judaea, became a renowned teacher and healer, and was crucified in Jerusalem on the orders of Pontius Pilate. Interestingly, the Gospels tell nothing about the life of Jesus from the age of twelve to thirty. Some writers have theorized that Jesus developed his spiritual powers and knowledge during this time by studying and practicing Tantra in India or Nepal. There is little evidence of this. However, there are many similarities between the teaching of Jesus as described in the gnostic Gospels and Buddhism.[2,3]

Although there is no question that Jesus was a great teacher of morality and offered a new and radical path of Judaism, the modern Church has accepted a version of his teachings that are from the three synoptic Gospels of Mark, Matthew, and Luke, plus the Gospel of John. The other Gospels were branded heretical—most notably the gnostic Gospels. How did this orthodox version of Christianity come about? Much of the credit for how the life and teachings of Jesus are viewed today falls on the Roman Emperor Constantine. He wanted to make Christianity the official religion of the Roman Empire, and toward this end, he called a council of orthodox Christian bishops in Nicaea in 325 AD. The Emperor ordered that the bishops remain in the castle until they came to a consensus. This they achieved and the resulting Church canon is the Nicene Creed.

This Council marked the beginning of Constantine's influence over the Church and gave the orthodox wing of Christendom official status within the Roman Empire. Gradually all other Christian sects were branded heretical and wiped out. However, an impartial, open-minded view of the literature would suggest that Jesus had achieved the state of seer or prophet and realized that he and the Father were the same. He refers to his connection with the Father repeatedly in the New Testament. It is unusual for someone to say they are God, but for a seer who experiences himself as one with the Cosmic Entity it is just as natural as saying "I exist." This is the essence of the Unity Principle, and we see that he had this understanding of the universe in many of his teachings.

Jesus talked often about reaching the kingdom of God. However, what is this kingdom and where does it lie? Is it the same as heaven? Jesus teaches that this kingdom lies within us, and that it is not heaven.

Once, having been asked by the Pharisees when the kingdom of God would come, Jesus replied:

> The kingdom of God does not come with your careful observation, nor will people say, 'Here it is,' or 'there it is,' because the kingdom of God is within you. (Luke 17: 20-21)

Again, in the gnostic Gospel of Thomas Jesus says:

> If those who lead you say to you, 'See, the kingdom is in the sky,' then the birds of the sky will precede you. If they say to you, 'It is in the sea,' then the fish will precede you. Rather, the kingdom is inside of you, and it is outside of you. When you come to know yourselves, then you will become known, and you will realize that it is you who are the sons of the living Father. But, if you

will not know yourselves, you dwell in poverty and it is you who are that poverty.

This passage reinforces the understanding that the gnostic Christians believed that knowledge of self was knowledge of God.

Deepak Chopra points out in his book *The Third Jesus* how these and many other passages from the Bible suggest that the kingdom of God is a state of consciousness—God-consciousness to be precise.[4] Hence, a strong argument can be made that Jesus taught the Unity Principle, but that the Church has mostly interpreted his words in such a way as to reinforce their agenda of duality, which has allowed the clerical hierarchy to exercise greater control over their flock.

Muhammad (570-632 AD)

Muhammad is regarded as the prophet and founder of Islam. He was born in Mecca, and after becoming discontented with his life, he went to the mountains to practice meditation. At the age of forty, he had his first revelation of God and soon afterward, he began preaching his version of monotheism. He taught that God is One and nothing less than complete surrender or devotion to God can lead to a place at his feet in heaven. According to Islam, Muhammad's revelations became the verses of the Quran.

Although Muhammad taught monotheism and morality, he preached a dualistic theology in which man was eternally separate from God. However, several Sufi saints developed mystical versions of Islam that were panentheistic. Ismaili Islam (Nazari Islam) affirms that God is omniscient, omnipotent, and omnipresent, and that everything in creation is formed from the vibrations that emanate from the Godhead. Hence, these paths of Islam are monistic and teach the Unity Principle that everything originates and is one within God.

Shankara (~790-820 AD)

Shankara lived during the ninth century in India during a time when Buddhism had become the dominant religion. He traveled throughout the country reuniting the fragmented Hindu religion under a systematic monistic philosophy based upon the Upanishads called Advaita Vedanta. Shankara was instrumental in reviving and unifying Hinduism in India. Now recognized as one of India's greatest mystics and teachers, Shankara taught that dualism was an illusion and that everything was the manifestation of Brahma; to know one's essence or atman, the core of oneself, is to know Brahma. His teaching that only God is truth and this world is false denied the importance of living in this world and may have contributed to a philosophy of indifference about the socio-economic plight of much of the Indian population.

Shrii Shrii Anandamurti (1922-1990)

Anandamurti was born Prabhat Ranjan Sarkar in Jamalpur, Bihar, India on the full moon in May 1922. From a very early age he exhibited many extraordinary abilities, such as practicing meditation by himself without a teacher; initiating much older persons in meditation; displaying great knowledge of languages, spiritual concepts, medicine, and various other topics, all gained without the help of teachers or books.

As a young student, Sarkar was considered by his teachers to be exceptionally bright but often bored with school and prone to daydreaming. At age seventeen, he left Jamalpur for Calcutta to attend the University of Calcutta but had to quit his studies two years later in order to support his family following the death of his father. For the next sixteen years, he worked as an accountant at the railway headquarters in his hometown of Jamalpur while teaching the spiritual practices of Tantric yoga to all the people that came to him for instruction.

In 1955, while still working in the railway office, Sarkar formed the organization Ananda Marga (Path of Bliss) with the twin purposes of spiritual progress and social change. He accepted the yogic name Anandamurti, which means "embodiment of bliss." He began training missionaries (acharyas and *avadhutas*) to spread his teachings of "self-realization and service to humanity" all over India and soon throughout the world. The Ananda Marga organization eventually grew to become a large and multi-faceted organization with members in over 130 countries, having different branches dedicated to the physical, psychic, and spiritual advancement of humanity.

Anandamurti became known as a spiritual teacher, scientist, philosopher, neo-humanist, social theorist, linguist, artist, and economist. He wrote over two hundred books on various subjects such as history, spirituality, sociology, education, Tantra, yoga, medicine, ethics, psychology, humanities, linguistics, economics, ecology, farming, music, and literature. For example, in the fields of music, literature, and art, Anandamurti founded RAWA (Renaissance Artists and Writers Association), which urged artists to produce art for service and blessedness rather than merely "art for art's sake," and gave guidelines for achieving this goal.

Although his most important contributions may have been in the arena of spirituality and the ancient science of Tantra yoga, Anandamurti was fluent in dozens of languages and wrote several volumes that explained the evolution of words and phrases from Sanskrit and other ancient languages into modern tongues. He also wrote children's stories, fiction, comedy, and drama. He proposed a new socio-economic system—PROUT (Progressive Utilization Theory)—as a humane alternative to communism and capitalism. His most impressive work was the over five thousand songs known as *Prabhat Samgiita* (Songs of the New Dawn) that he composed in the eight years prior to his death in October 1990. These beautiful and spiritually inspiring songs express in poetic form love for the Supreme Entity.

Anandamurti adjusted the ancient science of Tantra yoga to meet the needs of this age and taught a practical system of meditation and yoga for the all-around physical, mental, and spiritual development of the individual. The cosmology he taught was both rational and scientific and was based on the Unity Principle that everything is formed from the immanence of a transcendent God-consciousness.

Anandamurti's followers recognize him as a spiritually realized master and the third *mahasambhuti* to be born on this world following Shiva and Krishna. This is based on the following observations: he displayed omniscience; appeared to be fully realized from the time of birth and did not have a teacher; he could raise the kundalini of others to the *sahasrara* chakra, conferring on them *nirvikalpa* samadhi. His legacy will undoubtedly show that he had a tremendous impact on the economic, social, intellectual, and spiritual development of human civilization.

OTHER CONTEMPORARY PERSONALITIES

The word yoga is derived from the Sanskrit root *yuj*, meaning "to yoke" or "to unite." Yoga is a comprehensive system designed for the physical, mental, and spiritual upliftment of the individual. As implied by the name yoga, its philosophy is identical to that of the Unity Principle. Hence, some of the first persons to popularize the philosophy of the Unity Principle to the West were Indian yogis. The first was Vivekananda, who visited the United States in 1893; he was a disciple of the great yogic master Ramakrishna. Vivekananda was followed by Yogananda, who came to the United States in 1920, went on to found the Self-Realization Fellowship, and write the book *Autobiography of a Yogi*, which made a deep impression upon many people during the New Age culture of the sixties and seventies.

Chogyam Trungpa Rinpoche, a Buddhist meditation teacher, came to the United States in 1970 and founded a number of meditation centers, including the Naropa Institute and University in Boulder, Colorado. These eastern personalities were followed by various other swamis, gurus, maharishis, acharyas, *avadhutas*, Buddhists, Zen masters, etc., almost too numerous to mention.

Many other "home-grown" persons have delineated the Unity Principle in their teachings and writings. Rudolf Steiner called his spiritual science Anthroposophy. Mary Baker Eddy called it Christian Science, while Aldous Huxley called it the Perennial Philosophy. More recently many authors have introduced the idea of the Unity Principle in popular books—including the Dalai Lama, Deepak Chopra, and Eckhart Tolle.

13

The Unity Principle in Scripture

UPANISHADS

The Upanishads are Vedic scriptures that form the core teachings of Hinduism. The Upanishads teach the monism of the Unity Principle. A central theme of the Upanishads is "Thou art That"; that is, the self is identical to Brahma, the Cosmic Entity. Brahma is everything, is indeed this world. For example, the Isha Upanishad identifies God as present in everything:

> All this, which moves and changes, is covered by God, however insignificant it may be. Therefore, enjoy all with a sense of renunciation. Do not covet others' wealth, nobody steals anything—it is all God.

Again:

> Whoever sees all beings in the soul and the soul in all beings... What delusion or sorrow is there for one who sees unity? It has filled all. It is radiant, incorporeal, and invulnerable. Wise, intelligent, encompassing, self-existent, it organizes objects throughout eternity.

Again:

> There are not many but only One. Who sees variety and not the unity wanders from death to death.

In the Katha Upanishad, the boy Naciketa goes to Yama, the god of death, to learn intuitional science. Yama tells him:

> Beyond the senses are the objects (of the senses), and beyond the objects is the mind and understanding, and beyond the understanding is the great Self. Beyond the Great Self is the unmanifest; beyond the unmanifest is the spirit. Beyond the spirit, there is nothing. That is the end (of the journey); that is the final goal. Smaller than the small, greater than the great, the Self is set in the heart of every creature.
>
> The unstriving man beholds Him, freed from sorrow. Through tranquility of the mind and the senses (he sees) the greatness of the Self.

The Mundaka Upanishad states:

> The weak and timid cannot realize the Self. Self-realization is not possible through intellect or hearing spiritual discourse. One who welcomes God in every activity, through a thoroughly controlled and disciplined life, to him the Soul is revealed.
>
> The *Paramapurusha* is transcendental; he is formless. He is birthless, without breath (the vital life force) and without mind. He is pure and superior to everything.
>
> Just as the flowing rivers disappear in the ocean casting off name and shape, even so the knower, freed from name and shape, attains to the Divine Person, higher than the high.
>
> The knower of God becomes God.

Bhagavad Gita

As mentioned in Chapter 12 the Bhagavad Gita is from the Mahabharata. It is a story about the interaction between Krishna and his disciple Arjuna before the start of the Mahabharata war. In the story, Arjuna was feeling morally confused and hesitant to fight his friends and relatives. It was natural for Arjuna to feel compassion for his friends and relatives, and not wish to harm them, but Krishna points out that to uphold dharma is a just war, and that it is his duty as a warrior to fight. Krishna explains that Arjuna's dilemma is caused by his "I feeling" and has no validity in the Cosmic World.

Krishna exhorted Arjuna as follows:

> Oh, Arjuna, don't think that you are killing anyone. No one ever kills or is ever killed. Weapons can never kill anyone since the Self is immortal.

Krishna then displays his Cosmic Form to Arjuna and tells him:

> Things are already planned and predestined by Me. I have already planned everything out in My mind. You are just an instrument.

Hence, Krishna both taught and revealed to Arjuna the Unity Principle. Whatever takes place in this universe is designed by the Cosmic Will—everything takes place within the mind of the Cosmic Entity.

Taoist Scriptures

One of the classic books of Taoist literature is *The Book of Chuang Tzu*, which was written about 300 BCE. In this book, Chuang Tzu clearly delineates the Unity Principle:

> Do not ask whether the Principle is in this or in that; it is in all beings. It is on this account that we apply to it the epithets of supreme, universal, total. It has ordained that all things should be limited, but is itself unlimited, infinite. As to what pertains to manifestation, the Principle causes the succession of its phases, but is not this succession. It is the author of causes and effects, but is not the causes and effects. It is the author of condensations and dissipations (birth and death, changes of state), but is not Itself condensations and dissipations. All proceeds from it and is under Its influence. It is in all things

but is not identical with beings, for it is neither differentiated nor limited.[1]

Ananda Sutram

Ananda Sutram was written by Shrii Shrii Anandamurti and first published in 1962.[2] Sutras are aphorisms written in Sanskrit and designed to describe the nature of reality and how to obtain divine bliss. *Ananda Sutram* is perhaps the most concise description of the Unity Principle ever written. The first chapter of this scripture describes how Consciousness is transformed into Cosmic Mind, then into energy and matter, and finally into plant, animal, and human life forms. The second chapter deals with human dharma and the nature of the universe. The third chapter describes the nature of the cosmic and individual minds, and how spiritual practices allow one to overcome bondage. Chapter 4 details the Tantric theory of creation and the nature of the chakras and kundalini energy.

The last chapter describes the socio-economic system based on spiritual principles that Anandamurti calls PROUT. According to PROUT, a healthy socio-economic system must insure that the basic necessities of life are guaranteed for all and that the standard of living is progressively improved for the good of society. Avadhutika Ananda Mitra has written an excellent commentary on *Ananda Sutram*.[3]

14

Scientists on Unity

ALBERT EINSTEIN

Albert Einstein (1875-1955) is best known for his work on the theory of relativity and the nature of space, time, and gravity. However, Einstein also made major contributions in the realm of quantum physics. For example, Einstein received the Nobel Prize in 1921 for his work on the photoelectric effect—a quantum phenomenon. He also published papers on wave-particle duality, the quantum theory of atomic motion, zero-point energy, and the quantum theory of monatomic gases, which predicted accurately the existence of Bose-Einstein condensates.

Einstein appeared to have been a material realist who believed that there should be a one-to-one correspondence between physical reality and physical theory to explain that reality. Perhaps because of this bias, and because he believed that nothing in the physical universe could exceed the speed of light, he had a hard time accepting the quantum physics of Neils Bohr and his colleagues in Denmark (the Copenhagen Interpretation), which postulated that entangled quantum particles are in instantaneous communication with one another. Furthermore, the idea that there is uncertainty in every quantum event went against his sentiment that if there were a God, he would not create a universe governed by chance occurrences.

Although Einstein came up with many scientific objections to the quantum theory of Bohr, ultimately, his objections were shown to be without merit and the physics of Bohr and colleagues was proven correct. Near the end of his life, Einstein apparently realized that the nonlocality of time (his hypothesis) and space (Bohr's hypothesis) implied that there was a fundamental unity of all things, for he is quoted as saying:

> A human being is a part of the Whole, called by us the "Universe," a part limited in time and space. He experiences himself, his thoughts and feelings as something separate from the rest—a kind of optical illusion of his consciousness. This delusion is a kind of prison for us, restricting us to our personal desires and to affection from a few persons nearest to us. Our task must be to free ourselves from the prison by widening our circle of compassion to embrace all living creatures and the whole of nature in its beauty. Nobody is able to achieve this completely, but the striving for such achievement is in itself a part of the liberation and a foundation for inner security.[1]

In this quote, Einstein is exclaiming the basic concept of the Unity Principle, that is, that discreetness is a macroscopic illusion. Einstein also addresses liberation of the self or freeing oneself from the illusionary bondage of the personal ego:

> The true value of a human being is determined primarily by the measure and the sense in which he has attained to liberation from the self.[2]

Max Planck

Max Planck (1858-1947) was truly the father of quantum physics. Born and raised in Germany, he was the first scientist to postulate the idea that nature is not continuous but rather consists of discrete packets of energy or quanta. Based on his studies of the radiation emitted by a heated piece of metal (known as black-body radiation), he postulated that the electromagnetic radiation emitted by the body could only be a multiple of a constant, known as Planck's constant, times the frequency of the radiation. Planck's discovery ushered in the new era of quantum mechanics, and he may have been one of the first scientists to see the connection between the new physics of the quantum realm and its interplay with consciousness and the unity of all things. For example, he is quoted as saying:

I regard consciousness as fundamental. I regard matter as a derivative of consciousness. We cannot get behind consciousness. Everything we talk about, everything that we postulate as existing, requires consciousness.[3]

WOLFGANG PAULI

Wolfgang Pauli (1900-1958) was born in Austria and was one of the pioneers of quantum physics. He was a brilliant scientist and is best known for his discovery of spin theory and the Pauli exclusion principle that bears his name. This principle was important for understanding the structure of matter, and chemistry in particular.

Pauli was not your average empirical scientist. He believed that the universe had inherent order and that it could be best understood using intuition, or through experience of certain archetypes that exist in the collective unconscious. In this respect, his beliefs were similar to those of the seventeenth-century mathematician and astronomer Johannes Kepler and the modern psychologist Carl Jung.

Pauli had an understanding of the unity of all things that was reflected in his work on the unification of opposites. For example when it came to the complementary aspect of physical phenomena he wrote:

> The idea of complementarity in modern physics has demonstrated to us, in a new kind of synthesis, that the contradiction in the applications of old contrasting conceptions (such as particle and wave) is only apparent... The only acceptable point of view appears to be the one that recognizes both sides of reality—the quantitative and the qualitative, the physical and the psychical—as compatible with each other and can embrace them simultaneously.[4]

Pauli called for the use of the Unity Principle for obtaining an understanding of the "deeper invisible reality" that transcends our traditional macroscopic descriptions of reality.

> For I suspect that the alchemist's attempt at a unitary psychophysical language miscarried only because it was related to a visible concrete reality. But, in physics today, we have an invisible reality

(of atomic objects) in which the observer intervenes with a certain freedom (and is thereby confronted with the alternatives of "choice and sacrifice"); in the psychology of the unconscious we have processes which cannot always be unambiguously ascribed to a particular subject. The attempt at a psychophysical monism seems to me now essentially more promising, given that the relevant unitary language (unknown as yet, and neutral in regard to the psychophysical antithesis) would relate to a deeper invisible reality. We should then have found a mode of expression for the unity of all being, transcending the causality of classical physics as a form of correspondence; a unity of which the psychophysical interrelation, and the coincidence of a priori instinctive forms of ideation with external perceptions, are special cases. On such view, traditional ontology and metaphysics become the sacrifice, but the choice falls on the unity of being.[5]

Again, Pauli comments on the mystical unity of mind, body, and spirit in a letter to theologians:

I shall not venture to make predictions about the future. But, contrary to the strict division of the activity of the human spirit into separate departments—a division prevailing since the nineteenth century—I consider the ambition of overcoming opposites, including also a synthesis embracing both rational understanding and the mystical experience of unity, to be the mythos, spoken or unspoken, of our present day and age.[6]

James Jeans

Sir James Jeans (1877-1946) made many important contributions to quantum physics, dynamic theory of gases, radiation, and stellar evolution. In his book *The Mysterious Universe,* he outlined his belief that everything in the universe is entangled.

It is the same, I think, with other more technical concepts, typified by the "exclusion principle," which seem to imply a sort of "action-at-a-distance" in both space and time—as though every bit of the universe knew what other distant bits were doing, and acted accordingly. To my mind, the laws which nature obeys are less suggestive of those, which a machine obeys in its motion than those, which a musician obeys in writing a fugue or a poet in composing a sonnet. The motions of electrons and atoms do not resemble those of the parts of a locomotive so much as those of the dancers in cotillion.[7]

Again, he states the fundamental concept of the Unity Principle that the universe is the internal psychic thought projection of the Cosmic Entity:

If the universe is a universe of thought, then its creation must have been an act of thought. Indeed, the finiteness of time and space almost compel us, of themselves, to picture the creation as an act of thought; the determination of the constants such as the radius of the universe and the number of electrons it contained imply thought, whose richness is measured by the immensity of these quantities. Time and space, which form the setting for the thought, must have come into being as part of this act. Primitive cosmologies pictured a creator working in space and time, forging sun, moon, and stars out of already existent raw material. Modern scientific theory compels us to think of the creator as working outside time and space—which are part of his creation—just as the artist is outside his canvas. It accords with the conjecture of Augustine: *Non in tempore, se cum tempore, finxit Deus mundum* (God creates the universe not in time but along with time). Indeed, the doctrine dates back as far as Plato (time and the heavens came into being at the same instant…Such was the mind and thought of God in the creation of time).[8]

Jeans disagreed with the philosophy of Descartes that mind and matter were separate and not connected. If this were true then the mind could not possibly affect the body. Thus, Jeans agreed with the idealist philosophers such as Berkeley who argued that since mind and matter do in fact interact they must be of the same nature and that matter must arise from mind.

Biology, studying the connection between the earlier links of the chain A, B, C, D (thoughts, sensations, actions, results) seems to be moving toward the conclusion that these are all of the same general nature. This is occasionally stated in the specific form that, as biologists believe C, D to be mechanical and material, A, B must also be mechanical and material, but apparently there would be at least equal warrant for stating it in the form that A, B are mental, C, D must also be mental. Physical science, troubling little about C, D, proceeds directly to the far end of the chain; its business is to study the working of X, Y, Z. And as it seems to me, its conclusions suggest that the end links of the chain, whether we go to the cosmos as a Whole or to the innermost structure of the atom, are of the same nature as A, B.[9]

Erwin Schrödinger

Erwin Schrödinger (1887-1961) is best known for his pioneering work in quantum mechanics and especially for the wave equation that bears his name and for which he received the Nobel Prize in Physics in 1933. Similar to Carl Jung and his collective unconscious, Schrödinger believed in what he called the One Mind, and in the unity of all things. Schrödinger held that this idea was supported by science.

Schrödinger is also well known for his thought experiment, which has been dubbed Schrödinger's cat. In this scenario the behavior of a quantum particle is linked to the life or death of a macroscopic entity—namely a cat. The cat is put in a box along with a trace amount of radioactive substance that might omit a beta particle every hour or so, along with a Geiger counter that is linked to a hammer that will break a vial of poison. There is no way of knowing when the substance will decay and hence whether the cat is alive or dead without looking in the box. Quantum indeterminacy would require that there exist a superposition of states in which the cat is half-dead and half-alive until an observation is made. This thought experiment illustrates one of the oddities of quantum mechanics—namely that a system can exist in two states simultaneously until the superposition collapses during conscious observation.

 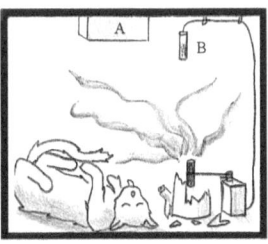

The paradox of Schrödinger's cat. If the radioactive substance in A emits a beta particle, then the Geiger counter at B will pick it up causing the hammer to break a bottle containing poison. Since, according to quantum mechanics, there is no way of knowing when the radioactive decay will take place until the box is opened and the cat observed, it must exist in both an alive and dead state simultaneously.

Schrödinger had a keen mystical insight into the functioning of the universe. Examples of his ideas on unity are taken from his books *My View of the World, Mind and Matter*, and *What Is Life?*

> Subject and object are only one. The barrier between them cannot be said to have broken down as a result of recent experience in the physical sciences, for this barrier does not exist.[10]

He considered the world to be the product of consciousness and the One Self the ultimate connection between individuals and the universe.

> To divide or multiply consciousness is something meaningless. In all the world, there is no kind of framework within which we can find consciousness in the plural; this is simply something we construct because of the spacio-temporal plurality of individuals, but it is a false construction. The categories of number, of whole and of parts are then simply not applicable to it; the most adequate

expression of the situation being this: the self-consciousness of the individual members are numerically identical both with one another and with that Self which they may be said to form at a higher level.[11]

His comments on mystical union are a statement of the Unity Principle.

> There is obviously only one alternative, namely the unification of minds or consciousness. Their multiplicity is only apparent; in truth, there is only one mind. This is the doctrine of the Upanishads, and not only of the Upanishads. The mystically experienced union with God regularly entails this attitude unless it is opposed by strong existing prejudices; this means that it is less easily accepted in the West than in the East.[12]

Schrödinger argued that our "I" is the "I" of God. He pointed out that our body functions as a pure mechanism according to the laws of nature; yet, we know by incontrovertible direct experience, that we are directing its actions. Hence, he argued that the "I" that controls the motion of atoms must be the "I" of God. Although this statement of the Unity Principle is strange from the standpoint of western thinking, he writes that it is seen consistently in the writing of mystics and is part and parcel of eastern philosophies.

Werner Heisenberg

Werner Heisenberg (1901-1976) is best known for his uncertainty principle of quantum theory, but he made numerous contributions to quantum physics and received the Nobel Prize in Physics in 1932. His understanding of quantum physics led him to express in several of his writings that separateness or the existence of distinct and nameable parts may be meaningless for the ultimate reality that physics seeks to deal with. For example, in his book *Across the Frontiers* he wrote:

> We know that there is an ever-changing variety of phenomena appearing to our senses. Yet we believe that ultimately it should be possible to trace them back somehow to some one principle.[13]

> The smallest units of matter are in fact no physical objects in the ordinary sense of the word; they are forms.[14]

In his book *Physics and Philosophy: The Revolution in Modern Science*, he writes:

> The atoms or the elementary particles…form a world of potentialities or possibilities rather than one of things or facts.[15]

He expresses the concept of unity at the quantum level and how the whole is formed from the intermingling of the parts.

> The elementary particles are certainly not eternal and indestructible units of matter, they can actually be transformed into each other.[16]

> The world thus appears as a complicated tissue of events, in which connections of different kinds alternate or overlap or combine and thereby determine the texture of the whole.[17]

DAVID BOHM

David Bohm (1917-1992) was born in the United States and made many important contributions to theoretical physics. In his book: *Wholeness and the Implicate Order*, he argues extensively that modern scientific theory implies that the universe displays indivisible wholeness.

> Ultimately, the entire universe (with all its particles, including those constituting human beings, their laboratories, observing instruments, etc.) has to be understood as a single undivided whole, in which analysis into separately and independently existent parts has no fundamental status.[18]

> What is needed is for man to give attention to his habit of fragmentary thought, to be aware of it, and thus bring it to an end. Man's approach to reality may then be whole, and so the response will be whole.[19]

What we perceive through the senses as empty space…is the ground for the existence of everything, including ourselves. The things that appear to our senses are derivative forms and their true meaning can be seen only when we consider the plenum, in which they are generated and sustained, and into which they must ultimately vanish.[20]

Quantum mechanics suggests that this is the way that phenomenal reality comes about from a deeper order in which it is enfolded. Reality unfolds to produce the visible order and folds back in. It is constantly unfolding and enfolding.[21]

To be confused about what is different and what is not, is to be confused about everything.[22]

Deep down the consciousness of mankind is one. This is a virtual certainty because even in the vacuum matter is one; and if we don't see this it's because we are blinding ourselves to it.[23]

Henry Margenau

Henry Margenau (1901-1997) was a German-born US-educated physicist and philosopher. His greatest contribution was not in the field of quantum physics but in integrating science and religion. His most important work in this field was his book The *Miracle of Existence*. Margenau argued that science as well as simple observation points to an undivided wholeness in the universe, which he called "Universal Mind." The concept of separateness or discreteness does not apply to the quantum level for matter, and similarly cannot be applied to the mental or superconscious levels. A mind without constituent parts is the Universal Mind, which is similar to the collective mind of Jung and Schrödinger or the Cosmic Mind, Paramatman, Brahma, God, etc.

Margenau argued that the fact that different living entities all perceive the same world despite differences in their brains is evidence for Universal Mind. Considering that we know the world only through our senses and brains, both of which have great differences, it is remarkable that everyone perceives the same picture of the world. He argues that this is only possible because we share the same One Mind. It is only because we all share the same consciousness that we perceive things the same way. If this were not true then there would be many different perceptions of reality.

> If my conclusions are correct, each individual is part of God or part of the Universal Mind. I use the phrase "part of" with hesitation, recalling its looseness and inapplicability even in recent physics. Perhaps a better way to put the matter is to say that each of us is the Universal Mind but inflicted with limitations that obscure all but a tiny fraction of its aspects and properties.[24]

Margenau believed that mind was nonmaterial and could function separately from the brain but was capable of influencing the brain on the quantum level, which would not require the mind to expend any energy in the process. He points out that conservation of energy as normally understood does not always apply on the quantum level—for example electrons can pass through barriers without expending any energy and particles with mass can be created out of the void of space. In the mind-body interaction, the brain can supply all the energy as mind causes a collapse of the wave function resulting in a particular probability event to take place in the physical organ.

> In very complicated physical systems such as the brain, the neurons, and sense organs, whose constituents are small enough to be governed by probabilistic quantum laws, the physical organ is always poised for a multitude of possible changes, each with a definite probability; if one change takes place that requires energy then the intricate organism furnishes it automatically. Hence, even if the mind has anything to do with the change, that is, if there is mind-body interaction, the mind would not be called on to furnish energy.[25]

If each individual mind is merely an illusionary part of Universal Mind, then why do we feel separate and bound by time, place, and person? Margenau would argue that it is our ego or personal wall that we build up from the time of childhood that reinforces our separateness from one another and from the One of Universal Mind, which is also responsible for the creation of the physical world. He believed that these limitations are overcome by the mystic who feels "at one" with the Supreme Entity.

The written works of Schrödinger, Bohm, and Margenau stand as some of the finest scientific arguments for the Unity Principle. Numerous other scientists have come after them bringing new and important arguments in favor of the idea that modern scientific discoveries and theories support the proposition that there exists a unity of all things.

FRITJOF CAPRA

Fritjof Capra is an Austrian-born American physicist who wrote the pioneering work *The Tao of Physics* about the parallels between modern physics and eastern mysticism. He has also written four other best-selling books: *The Turning Point: Science, Society, and the Rising Culture; Uncommon Wisdom; The Web of Life;* and *The Hidden Connections: A Science for Sustainable Living.*
The Tao of Physics was first published by a small New Age publisher but became so popular in the mid-seventies that it was picked up by a major US publishing house and has since been published in forty-three editions in twenty-three languages. Capra makes a persuasive argument that both eastern mysticism and modern physics point to an undivided wholeness or unity in the cosmos.

> The most important characteristic of the eastern worldview—one could almost say the essence of it—is the awareness of the unity and mutual interrelation of all things and events, the experience of all phenomena in the world as manifestations of a basic oneness. All things are seen as interdependent and inseparable parts of this Cosmic Whole, as different manifestations of the same Ultimate Reality.[26]

In regards to physics, he wrote:

> The basic oneness of the universe is not only the central characteristic of the mystical experience, but is also one of the most important revelations of modern physics. It becomes apparent at the atomic level and manifests itself more and more as one penetrates deeper into matter, down into the realm of subatomic particles...that the constituents of matter and the basic phenomena involving them

are all interconnected, interrelated and interdependent; that they cannot be understood as isolated entities, but only as integrated parts of the Whole.[27]

Capra points out that space, time, energy, and matter are all relative terms and are involved in a constant cosmic dance in which they are transformed into one another and into the void of space.

> The high-energy scattering experiments of the past decades have shown us the dynamic and ever changing nature of the particle world in the most striking way. Matter has appeared in these experiments as completely mutable. All particles can be transmuted into other particles; they can be created from energy and can vanish into energy. In this world, classical concepts like "elementary particle," "material substance," or "isolated object" have lost their meaning; the whole universe appears as a dynamic web of inseparable energy patterns.[28]

Fritjof Capra popularized the idea that modern physics supports and is consistent with the Unity Principle. Several other more recent books have been published relating science and spirituality. They begin where Capra left off. For example, physicists Menas Kafatos and Robert Nadeau argue in their book *The Conscious Universe: Part and Whole in Modern Physical Theory*, that Bell's theorem and the Aspect experiment demonstrate conclusively nonlocality on the quantum level that can only be explained by the fact that the universe is a Singularity or indivisible Whole.[29] They point out that for a universe that is One, whatever exists in the microcosm must exist in the macrocosm. Since we possess consciousness the universe is conscious and must be composed of and originate from Consciousness.

This is also the argument of physicist Amit Goswami in *The Self-Aware Universe: How Consciousness Creates the Material World*. Goswami points out that paradoxes of quantum mechanics and experiments that demonstrate nonlocality in time and space are best explained by the philosophy of monistic idealism. The universe is formed from the thought projection or transformed Consciousness of the Cosmic Entity. On the surface things may look different and separate to us, but in reality they are merely the manifestations of the One.[30]

Many other authors have made similar arguments relating scientific discoveries in the modern era to the philosophy of panentheism, namely that God is both indwelling and transcendent, that this creation consists of the internal thought projection of the Cosmic Entity.[31]

15

Western Attempts to Understand Unity

The Old School: God is Separate from His Creation

The Abrahamic religions of Judaism, Christianity, and Islam describe a God that is separate from his creation. This is the tenet of classical theism: God is the cause of the world, but he is not the material cause of the world. That is, the stuff of the creation is different from God, who is totally immaterial and perfect. God is spirit and the world is matter. God cannot be in the world of matter since he is perfect and the world is both imperfect and riddled with sin. Classic theism maintains that God is present and active in governing and organizing the universe but not "in" his creation except as spirit. The problem with this view of God is that it assumes that God can create something separate from himself, i.e. matter. However, if there were one thing God cannot do it would be to create another God. If he is truly an Infinite, All-encompassing Entity, then it is illogical that he could create something outside his being. Nevertheless, the classical theist concept of God as separate from his creation teaches that matter arises from God, yet is separate and not contained within the Divine Being. The irrationality of this idea about God and his creation has led many western thinkers to reject theism in favor of panentheism. In the panentheist concept of God, he is both the physical universe, or "in" his creation, and transcendent. In this way, panentheism differs from pantheism, which is the belief that the creation is identical to God.

Panentheistic Concepts of God and Creation

Eastern philosophers and prophets have always assumed that matter like everything else must be contained within the Divine Being. Hence, the eastern religions are panentheistic: God, the Transcendental Entity, is one with his creation and continuously manifests as the creation.

The important Christian philosopher and theologian St. Thomas Aquinas saw creation as an on-going process. Therefore, in some sense he proposed that God is continuously "in" his creation rather than an interested bystander. In other words, God is both transcendent and immanent. However, Aquinas believed God was perfect; his relationship with his created beings allowed for them to be autonomous and hence separate from him. Hence, Aquinas deviated in some respects from classical theism but did not fully embrace the concept of panentheism.

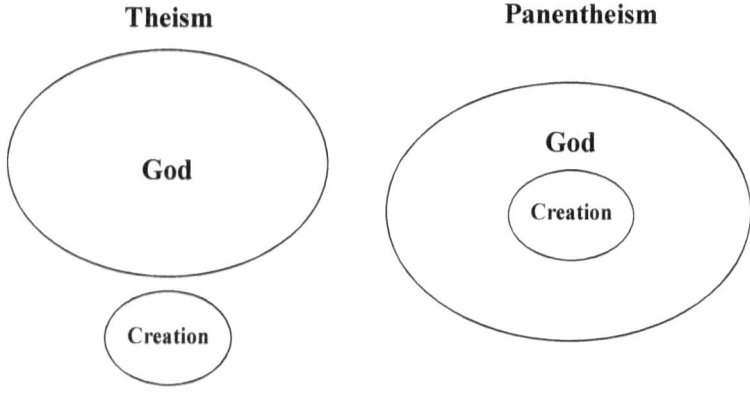

Diagrams depicting the creation as separate from God (theism) and the creation within God (panentheism).

One of the first westerners to touch on the concept of panentheism was the seventeenth-century Jewish philosopher Baruch Spinoza. Spinoza envisioned God as one with his creation. This alone would make Spinoza a pantheist (God = universe), but he also insisted that no attribute or substance can be truly divided, and hence God, the totality of everything, had to be above and beyond anything in nature. The One Ultimate Substance was God, but this Infinite Expression was manifest in many ways or modes in the created world. This so-called "dual aspect monism" of Spinoza is somewhat similar to the concept that the Infinite Entity, Brahma, expresses itself through the action of the Qualifying Principle, *Prakriti*. Although Spinoza saw the world as perfect because it was created from a perfect being, he had difficulty in accounting for both individual beings with their imperfections and a transcendent unified sense of self.

The pantheism of Spinoza was refined by other European philosophers, including Kant, Fichte, Schelling, and Hegel, who attempted to improve upon Spinoza's metaphysics by adding the idea of an active unifying principle of self. In particular, the nineteenth-century German philosopher Georg W. F. Hegel suggested that God was infinite and as such, nothing could be excluded from him. If God did not include the universe, then there would be an entity that was separate from him, and hence he would have boundaries and not be infinite. Hegelian philosophy is panentheistic since the infinite God is both the physical universe and the transcendental unifying spirit above and beyond the world. However, Hegel's philosophy was not monistic. Hegel considered himself a Christian and subscribed to many of its doctrines such as the description of God as the Father, Son, and Holy Spirit. Hegel also taught that time was a reality only for the emergence of God as "Being-in-and-for-itself." However, within the complete reality of God all time exists in a manner similar to block time.

Many western philosophers have followed in the footsteps of Spinoza and Hegel to expand and refine different varieties of panentheism. This list includes Whitehead, Krause, Leibniz, Hartshorne, and several other contemporary authors.[1] Modern scientific discoveries and thinking are clearly more in tune with the panentheistic concept that creation is an ongoing process and that if God truly exists then he must be an infinite entity and the finite world of creation must be within his being. For example, the Big Bang theory of cosmic evolution indicates that creation of the cosmos is an unfolding phenomenon, and Darwinism indicates that the same is true for the evolution of living organisms. Hence, the traditional concept of theism, that God is separate from his creation, that he brought the world into existence at a certain point in time with all its laws and forces to run autonomously, does not sit well with modern theological thinking.

Like Spinoza and Hegel before them, most western panentheists have come from the Christian tradition and often lack the unqualified monism of the Unity Principle. Dualistic panentheism attempts to account for the ideas of personal freedom or "free will," good and evil, our inherent individuality and eternal separateness from God, and Christ, the Son of God, as a being separate from the Father. Nor are all eastern panentheists free from dualism. For example, the Vaishnava Hindu Ramanuja and later A. C. Bhaktivedanta (of the Hare Krishna movement) taught that the material world arises from Brahma and is his maya or illusory joyful play. Once the individual soul comes to realize that it is not separate from the Supreme Entity, it is released from the cycle of reincarnation in the material world and lives eternally in paradise with full awareness that it is part of the body of God, though separate from him. As such, one delights in being a servant of the Lord.

It is this author's opinion that only monistic panentheism fits within the framework of modern science, spirituality, and the Unity Principle. Monistic panentheism should be understood to mean that the material universe is created from Consciousness and is God in his qualified state or *Saguna* Brahma. However, the unqualified aspect of Supreme Consciousness, or *Nirguna* Brahma, is beyond the manifest aspect of God. Qualified liberation or mukti comes when the individual "I" is merged in the Cosmic "I" of *Saguna* Brahma. However, in this exalted state of consciousness in which there is the experience of "I am God" there is still a tiny bondage since *Saguna* Brahma comes within the scope of *Prakriti* and in theory has infinite samskaras. Rather the spiritual aspirant upon death should seek to surrender their "I feeling" and merge into the realm of the Unqualified Supreme Consciousness. This is called moksha in Sanskrit and the experience is one of God as God.

16

The Mystical Vision of Unity

WHAT IS MYSTICISM?

Mysticism is the never-ending endeavor to find a link between the finite and the Infinite. In all humans, there is the urge to become one with the cosmos; thus, mysticism is the fundamental goal of spirituality, the experience, or feeling of being one with the cosmos. It is for this attainment that we have been born. It is the fulfillment of our being, the meaning of our existence, the purpose that underlies everything we desire and every action we take. In the Bhagavad Gita, Lord Krishna described mysticism in the following way:

> I appear before a person according to his or her desires. His or her whole being will be filled with My being. All the *jivas* (units) of this universe are rushing toward Me, knowingly or unknowingly. This is the final secret of the universe.

The mystical state of awareness is sometimes called Cosmic Consciousness. It is called a fourth state of consciousness in which one experiences their unity with all things. This indescribable state of awareness is as different from normal waking consciousness as the dream state is from normal consciousness.

Most animals possess simple consciousness and are aware of their bodies and their surroundings. Human beings possess simple consciousness as well as self-consciousness. We are aware of ourselves as separate from other entities and from our environment. We are mentally able to stand outside ourselves and make judgments about the operation of our mind. In other words, we are aware that the various mental states associated with our "I feeling" are separate and distinct from other people's "I feeling" and from everything else around us. We know that we know. This state of consciousness appears to be distinctly human and results from the preponderance of *Mahattattva* or "I feeling" that humans possess.

Rarely humans experience the fourth state of consciousness—i.e. Cosmic Consciousness in which the separation of the "I" from the universe that exists in self-consciousness breaks down and the "I" is experienced as "I am God." The mystical vision of unity may come as a flash or last for hours or days. As mentioned earlier it is sometimes called awakening, illumination, enlightenment, satori, samadhi, etc. Those who have experienced it compare it to waking up from a dream and realizing that the dream state was not real. The experience is profound and life changing.

This highest state of consciousness reveals that the universe is alive with the pulse of consciousness; it is ever-changing yet never-changing; everything is the Good. Evil is nonexistent; the concept of death is an absurdity; the universe is the Cosmic Entity and the Cosmic Entity is the universe; and it is every human being's birthright to experience the indescribable ecstasy of union with God.

Many people have had a true mystical experience and have tried to describe this ineffable state of consciousness. Despite differences in language, culture, religion, and era their feeble attempts at describing this state are remarkably similar—they all affirm the Unity Principle that everything is a manifestation of the One.

MYSTICS THROUGHOUT THE AGES

All the great religious prophets undoubtedly experienced a mystical connection with the Cosmic Entity. Shiva, Krishna, Moses, Buddha, Lao Tzu, Jesus, and Anandamurti all displayed the illumination characteristic of the mystical experience. Numerous other saints and sages have been described as mystics. These include St. Paul, Rumi, St. Catherine of Siena, Kabir, Mohammed, Hakuin, St. John of the Cross, St. Francis, Baal Shem Tov, Baha'u'llah, Ramakrishna, Ramana Maharshi, Swami Ramdas, Yogananda, Meher Baba, Krishnamurti, and Gopi Krishna.[1]

Richard Bucke argued that many historical figures had a mystical revelation of the unity of all things in his classic study of Cosmic Consciousness.[2]

Some of those that he included were Francis Bacon, William Blake, Walt Whitman, Edward Carpenter, William Wordsworth, Ralph Waldo Emerson, and Henry David Thoreau. Such historical figures probably experienced Cosmic Consciousness at some point in their life based upon their written descriptions of their experience, or as evidenced by the life-changing event that led to their moral elevation, the sudden change in their attitude about life and death, and their tremendous optimism and charisma.

Mystical illumination implies a transcendence of ego—a fundamental shift of consciousness from the individual self to the Whole or One. Concurrent with this shift comes a sense of timelessness and infinite spaciousness in which everything seems to be within oneself instead of without. The bondages of personality dissolve and there is a complete surrender of the individual will to the will of the Cosmic Entity.

People who have had such an experience agree that the experience is very difficult to describe in words. Their descriptions may take on different colors depending on the person's individual sentiments, religion, birthplace, culture, etc., but several similarities remain. One is the feeling of indescribable rapture, love, or ecstasy that accompanies their union with God. Secondly, people describe the experience as being more real, having greater clarity or vividness than ordinary waking consciousness—nothing like a dream or hallucination. Thirdly, the experience permanently changes people's attitude about death and the meaning and purpose of life. Hence, a transformative experience that lasts a lifetime. Below are descriptions of the mystical experience of a few individuals.

St. Catherine of Siena (1347-1380)

At the age of sixteen, Catherine joined the St. Dominic nunnery in Siena, Italy and developed a gift for contemplation of the Lord. She frequently experienced ecstatic union with God. One such rapture is described in her biography.

> On the feast day of the Apostle Paul's conversion Catherine was rapt in ecstasy and her spirit ascended so high that for three days

and nights she gave not the slightest sign of life. Those present believed her dead, or on the point of death. A few, however, who understood what was happening considered that she had been taken up by the Apostle into the third heaven. Time passed, and the ecstasy ended; but her spirit, drunk with the heavenly things it had seen, seemed so reluctant to return to the things of earth that she remained in a sort of daze, like a drunkard who is stupefied but not asleep.[3]

When asked about what she experienced during her mystical voyages she answered:

> Father, my soul saw and understood everything in the other world that to us is invisible: that is to say, the glory of the Saints and the pains of sinners. I have already told you: the memory cannot keep anything of it and words are not adequate to describe it; but as far as I can I will try to tell you about it. You can be certain, then, that my soul contemplated the Divine Essence; that is why I am now always so discontented with being in the prison of the body.[4]

RAMAKRISHNA (1836-1886)

Ramakrishna was an Indian sage, mystic, and practitioner and teacher of Tantric yoga. He was well known for his ability to enter the state of samadhi at will and sometimes even during everyday activities. He once described his mystical revelation of Truth.

> I do see the Supreme Being as the veritable Reality with my very eyes! Why then should I reason? I do actually see that it is the Absolute who has become all things around us; it is he who appears as the finite soul and the phenomenal world! One must have an awakening of the spirit within to see this reality. As long as one is unable to see him as the one reality, one must reason or discriminate saying, "Not this; Not this." Of course, it would not do for one merely to say, "I have seen beyond the possibility of a doubt that it is he who has become all!" Mere saying is not enough. By

the Lord's grace, the spirit must be quickened. Spiritual awakening is followed by samadhi. In this state, one forgets that one has a body; one loses all attachment to the things of the world... The spirit within being awakened, the next step is the realization of the Universal Spirit. It is the spirit (atman) that can realize the Spirit (Paramatman).[5]

ROBERT ADAMS (1928-1997)

Robert Adams was born and educated in the United States. At the age of fourteen, he had his first mystical experience, which forever changed his life, and by the age of sixteen, he began an earnest quest to find his spiritual teacher.

> When I had my spiritual awakening, I was fourteen years old. This body was sitting in a classroom taking a math test. And all of a sudden, I felt myself expanding. I never left my body, which proves that the body never existed to begin with. I felt the body expanding, and a brilliant light began to come out of my heart. I happened to see this light in all directions. I had peripheral vision, and this light was really my Self. It was not my body and brighter and brighter and brighter, the light of a thousand suns. I thought I would be burnt to a crisp, but alas, I wasn't. But, this brilliant light, which I was the center and also the circumference, expanded throughout the universe, and I was able to feel the planets, the stars, the galaxies, as myself. And, this light shone so bright, yet it was beautiful, it was bliss, it was ineffable, indescribable. After a while, the light began to fade away and there was no darkness. There was just a place between light and darkness, the place beyond the light. You can call it the void, but it wasn't just a void. It was this pure awareness I always talk about. I was aware that I AM THAT I AM. I was aware of the whole universe at the same time. There was no time, there was no space, there was just I am.[6]

Gopi Krishna (1903–1984)

Gopi Krishna was a yogi mystic born in India who began practicing meditation at the age of seventeen. He wrote about his mystical experiences, which he attributed to the rising of kundalini energy, in his autobiography, *Living with Kundalini*, beginning with his first such experience.

> Suddenly, with a roar like that of a waterfall, I felt a stream of liquid light entering my brain through the spinal cord. Entirely unprepared for such a development, I was completely taken by surprise; but regaining self-control instantaneously, I remained sitting in the same posture, keeping my mind on the point of concentration. The illumination grew brighter and brighter, the roaring louder, I experienced a rocking sensation and then felt myself slipping out of my body, entirely enveloped in a halo of light. It is impossible to describe the experience accurately. I felt the point of consciousness that was myself growing wider surrounded by waves of light. It grew wider and wider, spreading outward while the body, normally the immediate object of its perception, appeared to have receded into the distance until I became entirely unconscious of it. I was now all consciousness without any outline, without any idea of corporeal appendage, without any feeling or sensation coming from the senses, immersed in a sea of light simultaneously conscious and aware at every point, spread out, as it were, in all directions without any barrier or material obstruction. I was no longer myself, or to be more accurate, no longer as I knew myself to be, a small point of awareness confined to a body, but instead was a vast circle of consciousness in which the body was but a point, bathed in light and in a state of exultation and happiness impossible to describe.[7]

Gopi Krishna goes on to describe how the serpentine energy transformed his life creating many difficulties and wonders.

ECKHART TOLLE (1948–PRESENT)

Eckhart Tolle is best known for his best-selling books *The Power of Now* and *A New Earth*. In his first book, Tolle described the experiences that led to his awakening. He was thirty years old and suffering from continuous anxiety and periods of suicidal depression.

> One night I woke up feeling utterly despondent and longing for annihilation or nonexistence. But, suddenly I began to question who I was and why I felt so completely separated from myself. Am I one or two? If I cannot live with myself, there must be two of me: the 'I' and the 'self' that 'I' cannot live with. Maybe, I thought, "only one of them is real." I was so stunned by this strange realization that my mind stopped. I was fully conscious, but there were no more thoughts. Then I felt drawn into what seemed like a vortex of energy. It was a slow movement at first and then accelerated. I was gripped by an intense fear, and my body started to shake. I heard the words "resist nothing," as if spoken inside my chest. I could feel myself being sucked into a void. It felt as if the void was inside myself rather than outside. Suddenly, there was no more fear, and I let myself fall into the void. I have no recollection of what happened after that.[8]

Tolle's life changed forever when his false self-image or ego collapsed under the weight of his suffering. He described what remained as his true nature, the ever present "I am" or consciousness in its pure state without attachment to time, place or person (ego). This experience left him in a state of indescribable bliss for a period of almost two years.

AUTHOR'S EXPERIENCE

It was the summer of 1968. I was twenty-two years old and had just graduated from college. I got a summer job working for a pharmaceutical company near my home in Chicago before starting graduate school at Stanford University. A few months earlier I had picked up a book on yoga and began practicing some asanas and performing breathing exercises and simple meditation.

About the same time, I happened upon a book, *The Psychedelic Experience: A Manual Based on The Tibetan Book of the Dead* by Leary, Metzner, and Alpert.[9] This book stimulated my interest in psychedelic drugs. The book revealed a process by which one could take a psychedelic drug such as LSD and experience a journey into new realms of consciousness and transcend the ego state, followed by "rebirth" into a higher state of awareness. By simply taking one of these drugs, one could open the "doors of perception" as described by Huxley and have a transcendental experience.

Being trained as a chemist and aware of the chemical name for almost a dozen different hallucinogenic compounds I scanned the shelves of the chemical stockroom at my place of work and was surprised to find on a shelf a small bottle of methylenedioxyamphetamine (MDA). I "borrowed" a few milligrams of the liquid (enough for the standard dose), put it into an empty gelatin capsule, and took it home with the intention of experimenting later with the drug.

I prepared myself mentally and physically for my first psychedelic "trip" as described in the book. My goal was to experience the first bardo, whereupon the ego is annihilated and one experiences the pure essence or "I am" of being. One evening when I was alone at home and felt ready to take the drug, I skipped dinner and downed the capsule. I lit a candle and some incense and put on some sentient music. Then I sat comfortably in half-lotus posture and began my meditation. After about twenty minutes, I began to feel a wonderful sensation at the base of my spine that grew stronger and stronger and pulsated up my spine bringing me greater and greater bliss.

As this extremely pleasant sensation grew in intensity, I perceived the energy originating at the base of my spine gushing upward, flooding my brain with ecstatic warmth and light. These sensations were accompanied by a deep roar like that of the ocean, and then suddenly with a flash like lightning the energy came straight up my spine and lit up my brain in indescribable light. Every vibration I had been experiencing seemed to cease and I felt my being become nothing but infinite peace-space-bliss. It felt as though the entire universe was inside my being in a state of perfect peace and balance. The only thought in my mind was "I exist and am the One." I do not know how long I remained in this state of union, but the next morning I felt myself coming out of the trance and began to cry, having lost this state of blessedness. The next day it was impossible for me to work or do much of anything that required my attention to the crude world, and I continually fell into uncontrolled crying spells. The

THE MYSTICAL VISION OF UNITY 161

experience changed my life forever. From then on, I knew that getting back to that state was the only meaningful goal in life.

I took MDA and other psychedelic drugs like LSD on several subsequent occasions but never duplicated this experience. In fact, my insatiable thirst for this experience drove me down the road of continued drug use, despair, and thoughts of suicide. I eventually realized that drugs could not take me to where I wanted to go and I turned to spiritual teachers for help. I realize now that this initial experience in Cosmic Consciousness was the universe's way of opening my mind to the unfathomable potential that I had as a human being, but was not itself a path to enlightenment.

17

Unity — The Better Explanation

THE MULTIPLE PROBLEMS WITH MATERIAL REALISM

Material realism (MR) is simply the philosophical theory that the material universe is real and is the ultimate reality. It is based on the old concepts of classical physics that assumed that separate parts, taken together, constitute the whole of physical reality. In a sense the epistemology of MR is opposite to that of quantum mechanics, which theorizes that all parts are interrelated and hence parts of a whole.

We have pointed out that MR has many problems and deficiencies when it comes to explaining physical reality as revealed by modern scientific observations. Let us quickly review these deficiencies so that we can better understand why the Unity Principle offers a more rational and scientifically sound explanation for reality.

MR implies that cause follows effect (causal determinism), but this does not apply to quantum particles.

MR assumes that ultimately everything can be reduced to the properties of matter and energy. Mind and consciousness are a product of the brain. Therefore, the universe functions independent of our observation of it. On the contrary, the universe *does not* function independent of our observation of it. Measurements or observations *do* change the nature of physical reality. The mind of the experimenter affects and is an indispensable part of the experiment.

The theory of MR assumes that every particle in the universe exists independent of every other particle. There can be only local interactions between such particles. Yet experimental observations indicate that this assumption is also incorrect. Paired quantum particles such as photons or electrons behave as though they are "linked" or "entangled" even at great distances, and they influence one another simultaneously. Quantum objects seem to be spread out over space and time, like giant waves, or they exist in multiple energy levels simultaneously. They become fixed in space, time, and energy only after we observe or measure them. The established nonlocality of quantum particles negates the dogma of MR and suggests that the material universe is part of an undivided Whole or Singularity.

To account for nonlocality MR first proposed the theory of hidden variables. However, this theory was negated by Bell's theorem; therefore, MR theorists turned to the many-worlds hypothesis. According to this theory when a measurement is made we create another possible direction the world can take. In other words the world branches, and there are an almost infinite number of copies of the universe made every second; we exist in multiple universes simultaneously but have no knowledge of them.

MR theory assumes that space is real and that time is linear. However, experiments verifying relativity theory demonstrate that space may be bent by gravity, and time is entwined with space and is not linear.

To account for the extraordinary fine-tuning of physical forces that has allowed conscious beings to develop in the universe, material realists proposed multiverse theory and the anthropic principle. During the inflation phase of the Big Bang multiple universes were formed, each having slightly different rates of expansion and different values for the physical forces, particles, and constants. Out of the innumerable universes created, ours inflated at just the right speed so that all the matter that formed neither collapsed after a few million years nor became so spread out that stars and galaxies could never have formed.

By chance, ours was the universe that had a collection of physical forces that were perfectly balanced and were suitable for long-lived stars, galaxies, and planets to develop. In our universe, the physics and chemistry were just right for the evolution of living organisms. All this had to happen since we exist as living, conscious beings (anthropic principle). This is circular reasoning and violates the common sense idea that a simple explanation is better than a complex one.

MR fails to explain the development of extremely complex biological systems and structures by random mutations of DNA. Random mutations of DNA with selection of any resulting beneficial traits (material Darwinism) can explain why bacteria become resistant to certain antibiotics. However, it does not explain how complex structures have evolved in which many changes to DNA are required before any advantage is conferred upon an organism.

MR theory assumes that mind and consciousness are epiphenomena of matter. That is, mind and consciousness result from neurochemistry in the brain. However, if consciousness is a product of crude matter, then why are neuroscientists unable to find a localized structure in the brain responsible for it? And how can mind or consciousness affect matter by causing the quantum wave function to collapse in a certain way when a decision is made to make an observation one way and not another? In fact, according to quantum mechanics everything exists in a state of superimposed possibilities. Thus both the measuring device and, for example, the electrons being measured exist in this state; it follows that nothing physical can collapse the wave function. If this is true, then something nonphysical must be responsible—that thing being mind.

According to MR theory, the material universe is the ultimate reality, and therefore, God does not exist and mind is local and does not survive death. This means that there is no meaning to our pitiful existence that lasts for no more than a microsecond on the cosmic time scale. Whatever happens to us is mere happenstance or coincidental and there is no fruit to bear for our actions. There are no higher levels of mind, ESP is a myth, and religious and spiritual experiences are due to chemical imbalances in the brain and are not real. Our survival is most important, and our life is best spent seeking material gratification and pleasure. Self-esteem and power are more important than altruism and service. Why should we care about the welfare of those who do not fall into our little circle of friends and family since they are not connected to us? Should a theory that dismisses so much that makes us human be given much credibility? Humans experience altruism, love, self-awareness, intuition, creativity, mystical experiences, and a longing for the Infinite.

There is a huge body of scientific evidence demonstrating that humans possess ESP. Many people have described having out of body experiences. They have described with great accuracy events, conversations, and objects both near and far from their body while they seemingly lay unconscious. Even blind persons have described accurately scenes played out while their consciousness hovered outside their body.

Virtually any person willing to undergo regression hypnosis will remember living in one or more previous bodies. Such memories of past lives do not seem to be based on any events or persons they have met or read about in this life. In addition, studies of children who remember details of their previous life on earth indicate that they truly possess an intimate knowledge of a person who died before they were born.

Many people report having "touched" God in a religious or mystical experience. Such experiences are described as being more real than ordinary waking consciousness and the experience changes their life.

The skeptic-atheist-materialist (SAM) who subscribes to the theory of MR portrays all the studies and data supporting the reality of these experiences as flawed or faked. These experiences cannot be real since mind is an epiphenomenon of matter and therefore localized and dependent on the brain. This attitude is no different from that of a religious fundamentalist who rejects the idea that life evolved on this planet over billions of years because it does not fit into their belief system. Both the SAM and the religious fundamentalist seek to disparage the scientific data supporting the alternate hypothesis without even studying it.

Although materialism is a belief system that could best be described as myth, it still dominates much of western scientific and common thinking. Since its assumptions apply on the macro level, it appears to be a common sense approach to understanding the world. A new scientific paradigm based on holism is needed. However, most scientists are loath to embrace this. Why? In part, it may be due to a reluctance to give up classical physics in favor of the ambiguousness of quantum physics. Secondly, scientists are often ignorant that a logical alternative to MR exists. The western concept of God and religion may play a role. Traditional Christianity, Judaism, and Islam teach a dualistic concept of God, wherein God created a universe separate from himself. However, not all scientists are SAMs and many firmly believe in God but they may fear ridicule by associates if they go against the tide of MR. Personal and professional egos are at stake.

Clinging to the worn out paradigm of MR has harmed science. Complex and untestable theories (some say ridiculous) have been proposed in order to avoid the alternative explanation of reality—the Unity Principle. Almost no funding is available for research investigating the awesome potential of the human mind and consciousness.

In the end, perhaps the greatest failing of MR theory is that it is unable to provide any explanation for how the material that makes up the universe originated in the first place.

The Better Explanation for Reality—The Unity Principle

Perhaps the single most perplexing problem of modern physics is how measurement or observation can so radically change the nature of physical reality. The Unity Principle explains this and many other seemingly bizarre phenomena of quantum physics. For example, cause and effect, independence of interactions, separation of subject and object, and locality are not upheld because the universe is a Singularity. The one overriding characteristic of a Singularity is that every thing, every particle, every energy quantum, and every unit mind and consciousness is part of the One, dependent upon and connected to everything else.

Hence, cause and effect, and subject and object are complementary aspects of the same thing. Matter originates from Consciousness. Particles cannot act independently because they are intimately connected with one another. Locality is meaningless since the part is nothing but a complementary aspect of the One. What exists in the microcosm must also exist in the macrocosm and vice versa.

Everything in the universe seeks to be spread out or be nonlocal. Things become localized (the wave function collapses) only upon observation—with the involvement of mind. Hence, mind, which is nonmaterial, affects matter at the quantum level by collapsing the wave function into a specific state. We can willfully raise our arm because the nonmaterial mind can and does affect physical matter. In the final analysis, matter is simply a condensed form of energy, and energy is a condensed form of mind.

Observation affects the observed, and in reality, the observer and the observed are complementary aspects of the One. Consciousness influences matter precisely because matter is an epiphenomenon of consciousness. If consciousness were derived from matter, then logically it would be impossible for observation to affect physical reality.

The fact that uncertainty exists in the quantum realm is also consistent with the fact that matter is a product of consciousness. We can describe Cosmic Consciousness as a wave having infinite wavelength. Subsequent qualification of this wave by the Binding Principle, *Prakriti*, imparts a slight curvature to it and the "I feeling" or *Mahattattva* is born. Additional qualification or binding of the *Mahattattva* creates the *Ahamtattva* and the *Citta*, and from the *Citta* the five fundamental factors. Hence, the physical universe can be thought of as having a wave-like nature, and by definition, a wave is nonlocalized with indefinite position and speed.

Secondly, whether observed by a sense organ or measured by an instrument, a quantum particle cannot be pinned down. Sense organs and instruments have limited precision, and the very act of physically measuring or observing a quantum object affects it.

To the unit consciousness with its limited awareness, everything in the universe seems to function probabilistically and with uncertainty. Only the Cosmic Consciousness can defeat the imprecision inherent in measuring devices, sense organs, and unit minds; so in a sense God does not play with dice—it only appears that way to unit beings because we do not possess the omniscience of God.

The Unity Principle also explains the weird science of relativity. Space and time are actually complementary aspects of the same thing—space-time. Space-time can also be called ether. Although space-time is very subtle, it is nonetheless one of the five fundamental factors and possesses cruder qualities and properties that are absent in the subtler Cosmic Mind, which is both infinite and transcends space-time. For example, space has dimensions and is not infinite. The speed at which electromagnetic waves such as light or particles such as cosmic rays travel through space cannot exceed the speed of light. In addition, space emits a subtle vibration that is sometimes perceived as sound by persons who are highly attuned to such vibrations. Space is distorted or bent by a strong gravitational force. Space-time has locked within it tremendous energy and new particles are constantly emerging and then disappearing into it. These properties of space or ether prove that it is a fundamental substance even though it has no mass. Time is entwined with space and is affected by the speed that an object travels through space. As an object approaches the speed of light, space is compressed and the object's internal clock slows down. Time completely stops for electromagnetic waves or photons traveling at the speed of light. Hence, linear time is an illusion. This means that one cannot accurately describe the time it takes for an object to travel between two points in space. Space-time is four-dimensional and all time exists within its framework (block time).

Only events have meaning within the framework of space-time. A complete view of reality is afforded by having four-dimensional sight. Since humans lack such sight, we can only perceive reality from our particular perspective, which provides us with an incomplete vision of the Whole. Cosmic Consciousness has no such limitation; it has a complete view of all space and all events that have and will occur.

In classical physics, gravity is a force associated with objects that have mass. However, observations have demonstrated the validity of Einstein's general theory of relativity; they show that space-time is distorted or curved by gravity. Light traveling through curved space is also bent. This phenomenon is sometimes called "gravitational lensing." In a massive black hole, the curvature of space is so great that light cannot escape. Hence, the force of gravity affects and cannot be separated from space, time, and light. These entities are clearly entwined.

Gravity, the subtlest binding force, affects all five fundamental factors. The electroweak force affects the luminous factor and the strong force acts only in the nucleus of atoms. The interconnection between forces and the matter and energy that make up the physical universe is characteristic of an entangled reality. This is the hallmark of a universe in which there is unity of all things. The strange mechanics of relativity point to a universe that is a Singularity. Differentiation occurs only when a unit mind gets involved in the dance by the simple act of observation.

According to the Unity Principle, the subtler layers of unit mind are directly connected to the Cosmic Mind and are therefore nonlocal and beyond space-time. Extrasensory perception is a real phenomenon that occurs when the individual mind taps into the limitlessness and timelessness of the subtle layers of the unconscious or superconscious mind. These subtle layers reflect the Cosmic Mind, which is the storehouse of all knowledge, past, present, and future. The vast majority of people under normal circumstances have little or no connection with their superconscious mind, but studies on a large number of people or using a large number of repetitions for each subject have repeatedly demonstrated the validity of ESP.

The majority of people have had at least one experience during their lifetime in which they tapped into their superconscious mind. It might have been a religious or mystical experience, or an intuitive feeling that something was going to happen that turned out to be true. Sometimes it manifests as an awful feeling of dread that something terrible has happened to a loved one, which later turns out to be true. Other times it may be a strong feeling that one has been to a place before or met a person before when one knows that they have never been to that place or met that person. Most people know that there is more to the mind than the physical brain because they have had a nonlocal psychic experience.

For a very few individuals psychic abilities come naturally. Some of these people have been studied by scientists and have been shown to be able to accurately predict the future, know the past, diagnose illnesses, view distant events, etc. They seem to be able to connect to their unconscious mind and therefore to Cosmic Mind. Since the Cosmic Mind is subtler than ether or space-time, the Cosmic Mind has the four-dimensional sight needed for comprehending block time. Everything that has ever occurred or will occur within space-time is known to the Cosmic Entity. Persons who can gain a fleeting glimpse of the Infinite Mind of God can see future or past events.

Science now accepts the fact that the mind of an experimenter affects physical reality. Clearly, human beings possess the power to move a physical object like their hand with their mind. Therefore, it is no surprise that the mind can also affect physical objects separate from the body (telekinesis), as demonstrated by numerous experiments.

The unit mind and consciousness depend on the physical brain in order to function, but they are subtler than the body and survive physical death. This explains why people report having out-of-body experiences, particularly while under anesthesia, and while dying. As long as there is a thread-like attachment of the mind to the body, the mind can return to the body. If this cord is cut, however, death results and the bodiless mind with the help of the Cosmic Mind will seek out an embryo to begin the cycle of life again in a new body.

Reincarnation explains why so many children remember details of their previous lives. Thousands of case studies of people that remember their previous incarnations on earth indicate that the unit mind and consciousness survive physical death and find a home in a new body. Reincarnation also explains why some people are born with disabilities or other disadvantages while others seem to be born with every advantage.

The Unity Principle offers an alternative paradigm to materialism for the origin of the universe and the development of conscious beings that is more logical, complete, and consistent with scientific data. The Cycle of Creation explains how Consciousness is transformed into the Cosmic Mind and then into matter in the *saincara* phase of creation; and how living organisms evolve in the *prati-saincara* phase. As organisms develop through constant struggle for survival and attraction for the Great both their physical and mental structures become more complex and eventually sentient beings evolve that have a preponderance of "I feeling" or *Mahattattva*. Material Darwinism cannot satisfactorily explain the complex physical structures that make up our bodies, or why we possess self-awareness. Evolution is a fact of nature, but it is teleological or guided by the natural attraction of the Cosmic Entity. The endpoint for the evolution of species is union with the Supreme Entity.

The Unity Principle explains why there is exquisite beauty and harmony in the world. Human beings never cease to be awestruck by beautiful scenery or a colorful sunset. Our aesthetic sense results from our connection with the Infinite. This connection with the Whole lies at the core of our being. The connection is subtle, but under the right circumstances, we can feel overcome by the beauty of nature.

The Unity Principle explains animal instincts and homing behavior. Animals are guided by Cosmic Mind. In some ways their connection with the Cosmic Mind is stronger than ours is.

The fact that all the physical forces and laws governing the universe seem to be finely tuned to enable the formation of stars, galaxies, and the physics and chemistry needed to support life is explained by the fact that the material universe is created by and from Consciousness. Therefore, the universe reflects the infinite intelligence of the Cosmic Entity. To believe that consciousness evolved from random chaos is to ignore the evidence that the universe functions like a finely tuned and incredibly complex machine, a machine too complex to have been assembled merely by chance.

18

Practicing Unity

The Unity Principle implies that what is true above is also true below. In other words, the microcosm reflects the macrocosm. Ultimately, they are the same; however, we live in a relative reality that demands our attention and seems to have a reality of its own. There is a Sanskrit aphorism that describes this: *Brahma satyam jagadapi satyamapeksikam.* "Brahma is absolute truth and the universe is also truth but relative truth." Although Brahma is unchanging and infinite, when it comes under the influence of *Prakriti*, the Cosmic Entity comes into the realm of relativity under the bondages of time, place, and person. The microcosm or unit comes under the bondage of these three factors and is a relative entity that undergoes constant change.

The universe is also a relative entity and so the changeable world appears to be a relative truth to the changeable unit entity. Hence, we cannot completely escape this relative reality as long as we live in it and subscribe to it. To try to deny this world by calling it an illusion and saying only God is real is a mistake. We have to deal with this world in order to transcend the barrier separating us from the Cosmic Entity. However, what is this barrier separating us from the Cosmic Entity, why does it exist, and how can we overcome it?

The Ego Problem

Ego is difficult to overcome, but it must be dissolved if we are to change our way of looking at the universe and at one another. The sense of ego or separateness is essential in order to function in the relative reality of this world. A child needs to be able to discriminate between self and non-self. Gradually the ego develops and allows the child to make sense of and interact with the physical world. Without such development, it is not possible for the child to develop normally into adulthood. Ego is involved with all forms of action or doership and without it, we would be unable to function.

Throughout human evolution, threats to our survival have been met by an activation of the autonomic nervous system, which releases adrenaline and prepares us to fight or flee from danger. The self-preservation instinct that we share with animals was necessary for our survival as a species, but it is no longer as important for modern man. Nonetheless, we still tend to react to minor threats to our ego as if we were in danger of bodily harm. For example, someone criticizes us, hurts our feelings, or does not give us the respect we think we deserve; we instantly want to turn away from that person or shoot back with some biting remark of our own. The pain to our ego may last a long time and we may shun further contact with that person and say bad things about them to others.

The hurt that we feel when our ego is pricked is a result of our ignorance of the Unity Principle. As long as we subscribe to the illusion of duality—that we are separate and disconnected from everyone else—we cannot escape the constant clash that results from knocking our personal ego against those of other people and against the universe in general. Personal ego is boosted by achievements, such as wealth, fame, and power, and the more achievements attached to the personal ego the further one is led from the pure essence of the "I am" state of mind, which is our link to Absolute Truth.[1]

Ego awareness is trapped in the space-time of "me" and "mine" and is threatened by the idea of losing control or merging with all existence. The idea of holistic oneness requires a drastic revision to the materialistic model of reality. This is very frightening to the ego. For example, one materialist writes:

> I ask myself: do I really want to be one with the universe, so intimately intertwined with all of existence that my individual existence is meaningless? I find I much prefer the notion that I am a temporary bit of organized matter. At least I am my own bit of matter. Every thought and action that results from the remarkable interactions of my personal bag of atoms belongs to me alone. And so these

thoughts and actions carry far greater value than if they belonged to some Cosmic Mind that I cannot even dimly perceive. The mystical holist trades the real, pulsating life of the outer world for what he perceives as an inner world of peace. But, that peace is the peace of a prison. Science has always provided the means for breaking us free from the prisons of ignorance and superstition. I hope to convince you that science has not suddenly reversed its course and become yet another set of shackles for humanity to carry. On the contrary, science continues to provide the key that unlocks all of our chains so that our bodies and minds are free to roam the universe.[2]

In this passage the skeptic-atheist-materialist puts his faith in his body and ego and rejects the concept of a higher being or mind that would rob him of his freedom. The SAM calls upon science to answer all questions and free the mind from the "prison" of ignorance and superstition caused by belief in a Supreme Being. Implicit in the skeptic's doctrine is a fear of losing one's individuality by becoming one with the Cosmic Entity.[3]

This is the greatest fear of the ego—death. For in death, as in becoming One with the universe, there is an implied loss of control and individuality. Above all the ego wants to maintain and increase its control over the mind. Yet it is this ego that is the source of our separation from God, the source of our torments, pain, desires, anxiety, frustration, numbness, attachments, and isolation.

The ego can be identified with the *Ahamtattva* or "I do" feeling. Although it is a burden for realizing one's connection with the Cosmic Entity, as pointed out before, it is an indispensable component of our mind. For without the "I do" engine of *Ahamtattva* we would be like vegetables unable to make any movements or decisions. Therefore, the goal of the spiritualist is not to destroy the personal ego, but to powder it down and rely more and more upon the Cosmic *Ahamtattva* to perform action. Personal ego is the main barrier for perceiving the oneness that lies beyond.

Hence, personal ego has complementary aspects: on the one hand it allows us to function in the world and on the other it perpetuates the illusion of duality and individuality. One solution to the problem of ego would be to renounce this world as false or illusionary and retire to a cave in the mountains to practice sadhana. The Tantric path rejects this escapism.

Tantra recognizes that the relative world of duality is a reality that we must recognize and utilize in order to transcend our personal ego and draw nearer to understanding and experiencing the Absolute Reality. Instead of proclaiming that this world is an illusion, one must try to see every living organism and inanimate object as a manifestation of the Supreme Entity and serve those living organisms as though one were serving God. In this way, we can both powder down our personal ego and help bring other beings toward greater social, economic, and spiritual growth.

The practice of service and surrender will gradually destroy the personal ego, but we do not need to worry about becoming vegetables. Cosmic Ego (*Ahamtattva*) will fill the vacuum and over time, we will feel like we are acting according to the will of the Supreme Entity. This creates feelings of peace, harmony, and non-attachment from the results of our actions. Personal ego is like an umbrella shading the rays of the omnipresent, unchanging Self from our consciousness.

In reality, we are no more separate from the Source than rays of sunlight are separate from the sun. The ego blinds us and creates a false reality by obscuring the Ultimate Truth, covering it with layer upon layer of "I do this" and "I do that." Ego is the source of our arrogance and suffering; to believe that expanding its power should be the goal of our life is the epitome of ignorance. In short, we need to identify with the witness, not the performer. Unless and until we transform this self-centered component of our mind, it is impossible to make progress on the spiritual path to self-realization.

Western science and religion do little to invalidate the myths of separateness and material realism. As a society, we pay a high price for subscribing to these dogmas. The majority of ordinary people live by the principle of separateness. Living only on the lower planes of existence, they identify themselves with their bodies and lower minds. In this state of ignorance, they see themselves as separate from the world and from other human beings. They erect social and psychic barriers between themselves and others—barriers such as nationality, race, sex, religion, and economic status. All the endless conflicts that have occurred throughout history can be traced to ignorance of our true nature—unity. Feelings of separateness disconnect us from God, from one another, from other living organisms, from our world, and from the consequence of our actions.

Unity and the Human Condition

> The entire humankind of the universe constitutes one singular people. All humanity is bound together; those who are apt to

remain oblivious of this very simple truth, those who are prone to distort it, are the deadliest enemies of humanity. Today people should identify these foes very well and build up a healthy human society, totally ignoring all obstacles and difficulties. It must be borne in mind that so long as a magnificent, healthy and universalistic human society is not well established, humanity's entire culture, and civilization, its sacrifice, service and spiritual endeavor, shall not carry any worth whatsoever.[4]

Shrii Shrii Anandamurti
Ananda Vanii 1973

The purport of the Unity Principle is to look upon every person, every object of this universe as part of the One. Many have tried to jeopardize the unity of the human race by creating factions. Such persons have a vested interest in creating divisions; they survive on the mental weaknesses of people and on their dissensions. Such people are afraid of the spread of the ideology of holism and exhibit their intolerance toward it in all sorts of ways, such as creating divisiveness, false propaganda, and lies. However, knowledgeable people should not be influenced by the dogma of separateness; they must continually strive to perceive the unity of everything and everybody and act accordingly. Hindrances are beneficial to human beings on the path of unity and to fight against them results in mental and spiritual expansion.

As previously mentioned, human beings possess developed self-consciousness and are able to act in a self-directed manner. Hence, human beings are free to make choices about how they direct their minds. When human beings choose to ignore their underlying divinity and that of other persons and act in such a way as to harm themselves or other persons or living beings, then it is said that they act out of ignorance (*avidya*). When their actions reinforce their connection with the Infinite we say they act out of knowledge (*vidya*). However, all actions performed by the ego, whether mental or physical, cause reactions (samskaras) that may be experienced immediately or later. Our whole life can be considered an adventure designed to bring us closer to the Ultimate Oneness. We learn and grow by experiencing the fruit of our actions, whether good or bad. Our missteps bring us closer to God because we learn from them and are less apt to make the same mistakes in the future. Therefore, suffering has an important purpose. It forces us to look forward in the march toward perfection and cleanses us of the heavy debt that we accumulated by ignoring the fact that everything in the universe is connected, and the law that what goes around comes around.

Everything in the material world is formed from Consciousness and since human beings possess developed mind and consciousness, they have an underlying thirst for limitlessness. The attraction for the Great steers the unit mind toward the Cosmic Nucleus. This is the Source of all being and one must ultimately return to it in order to fulfill one's destiny. Hence, there is an underlying meaning and purpose to existence. Every human being is on an endless quest for the Great.

This may manifest as a religious or spiritual quest, or out of ignorance energies may be diverted to the material realm where they manifest as a quest for fame, power, or wealth. The desire to climb the highest mountain or be the greatest in one's sport is also born from our deep-seated connection with the Infinite. The thirst for limitlessness can be both our greatest boon and our greatest enemy. When it is quenched by spiritual practice, we feel blessed by God and experience bliss. When we try to quench its pull by activities of the lower mind or ego then inexorably there is pain and frustration. All sorts of mental and physical problems arise because of living only on the lower plane of existence. All the mental ills confronting individuals and societies today can be traced to ignorance of the Unity Principle.

When people finally realize that they are already on the path to perfection and desire to be guided by knowledge rather than by happenstance, they inevitably begin in earnest to perform some type of spiritual practice. For it is only with such practice that one enters the accelerated path of bliss. The practice of sadhana is nothing other than cosmic ideation, floating on the divine waves of bliss, drawing ever closer to the Nucleus of Creation. However, these practices are subtle in nature and individuals differ in their mental development and propensities. No one practice is suitable for all persons. For optimal results, the practice should be tailored to the individual like a fine suit of clothes. Therefore, sadhana is best learned from an accomplished teacher.

The mind normally jumps around like a monkey on amphetamines. When beginning a spiritual practice one may become more aware of the mind and its unruliness. One may think that the practice is making things worse. At this stage, it is often important to have the reassurance of a teacher or others experienced in meditation. After some time the beginner will experience greater and greater control over the mind and with it the occasional taste of peace and bliss. Finally, the practice becomes part of their daily routine and the meditator would rather miss a meal than miss a meditation session. As Anandamurti puts it:

> The movement and the path, the means and the chariot are all inseparably linked. The path is not always easily accessible, smooth and littered with flower petals; nor is it always inaccessible, thorny and covered with stones. One must keep one's eye fixed on the Goal. This Goal provides inspiration, supplies the means for forward movement, and makes the little lamps of life infinitely effulgent. Since eternity this very Goal has provided and is providing inspiration to all and will continue to do so in future; and by revitalizing the life-force as if with a flow of water, it will make the earth ever full of sweetness, and at the same time it will keep the triumphant flag of humanity flying on top of the golden mountain peak. So let one's vision be fixed on the Goal. There is no necessity to think of anything else.[5]
>
> Shrii Shrii Anandamurti
Ananda Vanii, 1987

Traditional Practices for Experiencing Unity

Enlightenment is awakening to a clear understanding of the unity that is. The word "awakening" is useful because it is analogous to awakening from a dream. While dreaming one does not question the reality of the images that are witnessed, even though they may be quite bizarre. Upon waking up one will immediately dismiss the dream as unreal. Similarly, the illusion of separateness or non-unity appears real to us, but upon awakening to an enlightened state, we know that this common state of consciousness was illusory.

In this awakened state, one loses the sense of a separate selfhood. It is identical to the state of consciousness of the seer. Everything is seen and felt as the One and there is an end to suffering. A natural state of bliss arises from the non-separateness that occurs when one lives totally in the here and now of "I am." One continues to experience life with its joys, pleasures, pain, and love, but these experiences are not resisted by the illusory "me." In other words, one's experiences are witnessed by the mind but do not affect in any way the "I that is."

Surrendering the ego is taught by most of the world's great religions in order to reduce people's sense of separation from God and to increase their love of God. For example, Jesus taught unconditional, self-sacrificing love for God and for all people. He preached service, humility, and forgiveness. He is quoted as saying that one should turn the other cheek if someone slaps you;[6] to love your enemies and pray for those that pursue, slander, and falsely challenge you.[7] Jesus said that one must become like a child to enter the kingdom of God, completely loving and trusting.[8] It is obvious that Jesus knew that the greatest barrier separating us from the Lord is our personal ego. His true teachings, if followed with real awareness, would greatly diminish the ego's control over our life and allow us to see through the illusion of separateness from God.

It is easy to talk or read about surrendering the ego and becoming detached from its worldly demands. It is another thing to do something about it so that we can achieve the unity we seek. The ego is tenacious. It disguises itself in desire and attachment. It will never give up control without a fight. On the other hand, the component of mind that lies above ego, the "I am," or *Mahattattva* is intimately connected to the One, but is merely a detached witness. How then can we utilize the higher consciousness of the Self to guide us or perform any action whatsoever?

The answer is that just as our genes have programmed the development of our body, the Cosmic Entity has programmed our psyche to long for the Infinite. We naturally experience some discontent or dissatisfaction with all things that are time-bound. If we listen to that inner voice, which tells us there has to be something more or better, then we are constantly reminded that there is more to life than the achievements of ego. Some people call that inner voice "conscience." Others call it the soul. In any case, it is a purely human condition that we have acquired by having a preponderance of the *Mahat* or the "I am" quality of mind.

That inner voice, which speaks to us in subtle whispers, is only quieted by the constant activity of the ego-mind or the stupor of unconsciousness, such as that brought on by the use of drugs. When we finally come to realize that there is more to life than the incessant quest for pleasurable experiences and achievements, then we can finally graduate to the level of seeker. As a seeker, we begin to listen to the message of that inner voice and utilize our ego-mind for powdering down the ego-mind. For example, we may practice surrender, service, non-attachment, *madhuvidya* (trying to see God in everything), and meditation. In each of these practices there is the "I" and there is some performance of that "I." Ego is involved in all activities, even those that bring us closer to our goal of unity.

Attempts to crush the ego by acts of self-flagellation, self-denial, self-denigration, or surrender to the will of other persons will succeed only in leading us away from our goal. Every act of surrender must be to that Cosmic Entity and service must be to that Entity without the thought of gaining recognition or thanks. Selfless service to the manifestations of the Supreme Consciousness serves to powder down the personal ego, and as a byproduct, service brings happiness to those who are served and to us.

Prayer can be of value for moving forward on the path to perfection. The lower forms of prayer in which one asks God for some favor such as "God, please let me become rich" or "God, do my enemies harm" have no value. God already knows everything about us; he knows exactly what we need. Selfish requests will not be honored. A higher form of prayer is to thank God for what we have. This too is mostly a waste of time if the thanks are in the hope of receiving some favor from him in return. We may feel that we are being humble by thanking God, but he does not need any thanks from us. He loves all of his created beings unconditionally.

The highest form of prayer is devotional prayer in which a person asks God for emancipation. This is different from seeking a favor from God, because God's purpose in creating humanity is to make the unit beings free like him. This is the wish of God, and everything in this creation is directed toward that end. Thus, the most valuable form of prayer is devotional prayer, which is also called bhakti. Since the mind takes on the qualities of the object of its concentration, if one constantly calls the Supreme Entity in their mind, they will indeed become like that Entity and eventually become filled with love for the Supreme.

The study of spiritual books and scriptures has proven effective for progressing on the path to unity. This can also be called *jinana* yoga. By such study, the intellect is stimulated and one may become filled with both exhilaration and humility. The mind can obtain concentration as well, but such study has limitations. Union with the Supreme Entity cannot be obtained by intellectual exercises alone; at best, one may become motivated to perform spiritual practices that will ultimately lead to unity.

Austerities can have value in moving forward on the path to unity. Examples include fasting, sexual continence, and silence. Austerities can strengthen the will and unlock spiritual energies that can be directed toward attaining unity. During long fasts, the mind naturally quiets down and concentration on the Supreme Entity becomes much easier. Austerities are not recommended for everyone and the downside is that one may develop pride in being able to perform them.

Performing work is the essence of karma yoga. We all have numerous samskaras that need to be burned before we can obtain the Supreme Beatitude. Action done with the idea that the fruits of the action are for my Lord rather than for my personal gain can burn off these reactive momenta without creating new ones. Physical work is healthy for both the mind and body. Selfless work to serve others (*seva*) is the highest form of karma yoga. There is a dual benefit: not only is social service rendered, but one's individual development is also enhanced.

Yoga asanas are helpful for attaining union with the Supreme. This practice is also called hatha yoga. These postures or asanas help in calming, stretching, and vitalizing the body and glands. Hatha yoga is popular in the West, but it should not be confused with real yoga. As we have seen, yoga is much more than physical exercises and postures. This does not mean that asanas lack value for attaining unity. A healthy body is important if one wishes to perform meditation. Besides asanas, hatha yoga prescribes a lacto-vegetarian diet free of tobacco, alcohol, and drugs. Such a diet is beneficial for the mind and body.

Yogis wanting to advance on the path to perfection have utilized breath control or pranayama for centuries. Pranayama is useful in purifying the *nadis* or psychic nerve channels. When these channels are purified, the latent spiritual energy (kundalini) can pass through the chakras and reach the higher centers in the brain. The practice of pranayama concentrates the mind and helps with meditation; it is also very beneficial for the physical body. Different pranayama techniques can also be used to cure diseases.

Kirtan (singing spiritual songs) and *japa* (repetition of a mantra or prayer) can be useful techniques for bringing about spiritual elevation. Sound is the subtlest of the five *tanmatras*, and when kirtan is sung the mind is elevated and at the same time concentrated on the object of the song—God. Hence, kirtan is valuable in preparing the mind for meditation. It also helps develop devotion for the Lord. It thus stands as one of the most valuable practices for countering the natural movement of the mind toward the crude and redirecting it toward the subtle. This after all is the goal of all spiritual practice: to direct the mind toward subtlety. *Japa* is somewhat similar to kirtan in that it entails repeating a name of the Lord or other mantra. However, the chant is repeated internally rather than externally. Like kirtan, the mind is directed toward the Supreme Entity and becomes subtler. Even a simple prayer can be used. For example, in the nineteenth century religious classic *The Way of the Pilgrim*, the narrator, a wandering hermit, attains a very high state of spiritual awareness by ceaselessly repeating the Jesus Prayer (Lord Jesus Christ, Son of God, have mercy on me, a sinner). However, repetition of a siddha (power) mantra is probably the most effective *japa* practice for advancing on the *vidya* path.

Satsang, or associating with other spiritual persons, may have an elevating effect upon one's psyche and that of other persons. Since the goal of unity is obtained when the mind becomes detached from crudity, it makes sense that one should try to associate with persons who are also trying to overcome crudity. The highest form of satsang is to have personal experience with a spiritual teacher or guru. This is called darshan in Sanskrit. The love radiated by a master guru results in elevating one's mind and creating feelings of peace, love, and awe. After such contact, aspirants often feel energized in their practices for attaining unity. Satsang in the form of group meditation is also beneficial for one's spiritual development. The collective vibration of the group can have a synergistic effect and enhance one's meditation.

Attending church or temple may elicit pious feelings and hence devotional sentiment. These institutional forms of worship can be useful in opening one's heart to the pervasive and unconditional love of the Divine Entity. Other forms of ritualistic religious expression such as Hindu puja may have a similar effect. However, all practices that rely on physical symbols and idols have limited value for attaining the Infinite Entity. At best, they can create pious feelings, some devotional sentiment, and create a desire to perform true spiritual practices.

Modern-day authors have also contributed to the list of spiritual techniques and practices. For example, Eckhart Tolle argues that being totally in the now leads to an awakened state since all suffering is caused by being attached to and worrying about the past or future—both of which are nonexistent or illusory.[9]

Deepak Chopra describes seven laws that lead to spiritual success or enlightenment.[10] He describes the physical universe as nothing other than the Self turning back within itself to experience itself as unit consciousness, mind, and physical matter. Moreover, the laws governing the universe are nothing but the whole process of Divinity becoming this universe. When we understand and apply these laws to our lives then we can attain our deepest desire—to become one with the Lord. In his book *The Way of the Wizard* Deepak Chopra describes twenty lessons for powdering down the personal ego, which he characterizes as the enemy that creates the illusion of separateness and robs us of the love, personal fulfillment, and connectedness that we long for as human beings.[11]

Other authors point out that conscious will is an illusion. Decisions are actually made in the subconscious (preconscious) mind and are determined by our inherited and neurological conditioning. Awakening occurs when one realizes that conscious will is an illusion and begins experiencing the here and now devoid of the illusion of a separate self.[12]

Zen Buddhism (Zazen) takes a somewhat similar path to enlightenment (satori). In one of its practices, the student is presented with an unsolvable problem known as a koan, such as "what is the sound of one hand clapping." The problem exists only in the mind. The minute the mind is suspended and sees through the illusion of an imaginary and separate self, the problem is solved. Separateness exists only in the mind—it is not real. The hand and the sound it makes are the same.

Some practitioners of Buddhism use meditation on nothingness in order to achieve enlightenment. In this practice, the meditator tries to empty their mind. If thoughts arise, they are allowed to pass with no attention put upon them. This is a difficult meditation technique that is usually done by more advanced practitioners, but it can be very valuable in calming the monkey-like mind and allowing one to become more proficient at being the witness, free from the illusion of acting.

Many writers such as Ram Dass suggest that we develop the witness.[13] To become a witness instead of a doer you can practice specific techniques and exercises that take you out of your normal patterns and allow you to observe the mind more clearly. Most of our actions and decisions are performed subconsciously. Any action performed with mindfulness, i.e. with full attention, can be used to power the witness. Mindfulness meditation practices are therefore useful for helping to strengthen the witness.

Surrender, Service, and *Madhuvidya*

Lord Krishna says that his maya, the force that creates confusion and distinctions, is very powerful, even insurmountable by human beings. "But those who surrender unto Me transcend these forces of Mine with My help." So according to this prophet, the illusion of duality caused by the bondage of personal ego is very powerful, but there is hope for people who seek God. Those who surrender to the Cosmic Entity can surmount the bondage of personal ego.

Personal ego is destroyed by surrendering to the will of the Supreme Consciousness, performing selfless service, and trying to see everybody and everything as manifestations of that Supreme Consciousness. These three activities are all related and are useful for piercing the illusion of duality that the personal ego creates. Let us look at these actions in more detail to see how we can apply them in our life.

Spiritual aspirants are told to try to surrender all their actions to Brahma so that they do not have to endure the reactions. This surrender is an important aspect of spiritual practice. The aspirant is taught to keep in mind that "This is That." The action is for God and by God. This is similar to the thought that we are acting according to his will, that he is the machine operator and we are the machine. By surrendering our will to his will, we are freed from the consequences of our action. Our inner voice or conscience will normally tell us if a proposed action is for selfish purposes. Actions that are done for personal gain or for selfish motives do not powder down the personal ego but tend to harden it. The consequence of such actions cannot be surrendered to the Supreme and will add more samskaric thread to the fabric of the personal ego.

Ultimately, the unity we seek is attained by surrender of our own individuality. To give up crude objects like money or real estate to a religious or spiritual group will not by itself bring us unity. If we want to attain the bliss of oneness, we must offer our own self. If we want to have the "Great I" we must give away our little "I." One must surrender completely and hold nothing back.

Such self-surrender is not akin to suicide. On the contrary, our soul will have its full expression. Our existence does not become contracted but enlarged since the Cosmic Ego will find expression through our mind and body instead of the personal ego. Unlike the personal ego, the Cosmic Ego is infinite in scope. Our ego can literally expand infinitely by surrendering it to the Cosmic Entity, and in the process, we will attain deep and lasting happiness.

True service is called *seva* in Sanskrit. True service is not a mutual interaction like business. True service is unilateral. You perform *seva* out of the goodness of your heart and require nothing in return, not even recognition or thanks. There are several types of *seva*. For example, one can give physical service such as help building a shelter, or medical service to the sick. One can perform security service such as protecting and helping the weak or economic service such as relief work, feeding the poor, or helping the needy find a job, etc.

A higher form of service is intellectual service, such as teaching skills, morality, and spiritual ideology and practices. Unlike the other forms of service the effect of intellectual service can be permanent in nature. Not that physical forms of service are less important—the effects are temporary in nature. The material needs of people and other living organisms must be satisfied to maintain life, and life must be maintained to reach God. All forms of true service weaken the grip of ego and bring one closer to the goal of unity. This occurs because selfless service is not performed by the personal ego. In such service, the ego is subjugated and one acts according to God's will. Selfless service can be a powerful tool for destroying the arrogance of the personal ego.

The highest form of service to God is spiritual service—performing sincere spiritual practices. Such practice helps one attain unity and helps others feel that unity through your example and psychic influence.

Madhuvidya, or ascribing godhood to every living organism and object, is a powerful spiritual practice. It powders down the personal ego. *Madhuvidya* is nothing but ideation upon the Supreme. The practice allows one to carry on a normal worldly life and yet through common experience draw closer to the Supreme Entity. For example, while feeding your daughter you might contemplate that you are not feeding your child but actually serving a manifestation of the Supreme Consciousness in the shape of your daughter. When watering your lawn you are doing a proper action to the manifestation of the Supreme Consciousness in the form of grass. And, when you are walking down the street and see other people one can try to envision them not as strangers but as manifestations of the Supreme.

Try to envision yourself as a wavelet in the Mind of God. At all times it is best to feel that he is watching us, aware of every thought and action we take. If one properly follows the cosmic ideation of *madhuvidya* then one will be set free from the bondages of actions. This practice can pervade one's exterior and interior with the ecstasy of cosmic bliss and permanently extinguish all afflictions.

Meditation, the Most Effective Practice

Many practices have evolved over the ages for developing an awakened state of consciousness. Of all the practices discussed, meditation or sadhana is probably the most powerful technique for achieving enlightenment. Meditation was discussed in Chapter 11, but it is useful to discuss why it is such a powerful technique for achieving unity.

The Supreme Entity sees everything—everything is within his mind. He is the Supreme Subject and everything else is his object. In other words, the Supreme Seer is the subject and the seen is the object. This Supreme Subjectivity is precisely what the meditator concentrates on as they attempt to pierce the veil of ego. Meditation quiets the restlessness of the monkey-like mind, but its purpose is not simply to produce peace and calm. The true purpose of meditation is to develop the ability to concentrate the mind, for it is only when the mind is fully concentrated on the Supreme Consciousness that the personal ego is transcended.

When all psychic tendencies and thought-waves are channeled toward the Supreme then the dynamic flow of the mind, concentrated at a single point, becomes one with the Supreme Entity. The finite microcosm merges in the macrocosm. The small "I" of the ego is lost in the infinite sea of the "Great I." In this state of total surrender, there is complete fulfillment; one realizes that individuality was actually an illusion as everything is continuously manifested by the Supreme. If one can surrender their all to the Supreme Entity by means of sadhana then they will attain liberation even while in their physical body.

Devotion: The Final Stage

The practices to obtain unity can be termed yoga. By yoga, we mean unification, not some physical exercises as often misconstrued in the West. We have pointed out that there are numerous types of yoga practice, but the highest form of yoga is bhakti yoga, the yoga of devotion.

Devotion is ideation upon and love of God. Bhakti is not so much a practice as a condition. It is the ultimate goal of all spiritual practices. To be ensconced in constant love and ideation of the Infinite Entity is to be a seer and see everything as God.

All great saints and prophets had an overwhelming love for God and they radiated this love. This naturally attracted many followers. Where there is love for the Infinite Entity there is no personal ego since ego is only involved with the attachments for finite things. Devotional love is directed inward as opposed to love for a spouse or child, which is directed outward.

Arousing and attaining love for God is not easy. Like Mother Teresa, one can spend their life tending to the poor, attending daily church services, reciting prayers or mantras, undergoing long fasts, or performing extreme acts of self-denial, and never obtain more than momentary feelings of devotion for the Lord. It helps to live by the Golden Rule and to try to love and respect all manifestations of the Lord; but devotion is a state of mind, not a practice. One must rely on spiritual practices such as surrender, service, *madhuvidya*, and most importantly meditation to obtain this state.

In order to obtain liberation, one must completely surrender one's body, mind, and soul to God. It is only by completely surrendering our little "I" that we can merge with and become one with the Cosmic Entity. To do this we have to have complete trust in the Supreme Entity. This trust comes from the devotional love we develop for the Supreme. Hence, there is a positive feedback loop: love for the Supreme begets trust, which begets surrender, which begets greater love, etc. Love of God is the necessary ingredient for starting the cycle, and therefore the goal of spiritual practice must be to develop bhakti.

The Unity Principle is summarized in the simple Sanskrit aphorism: *Baba Nam Kevalam,* which literally means that everything (*Kevalam*) results from the emanations (*Nama*) of the Cosmic Entity (Baba). I hope that the reader of this book has learned that although the Unity Principle provides a logical and satisfying explanation for the origin and purpose of the universe we live in, nonetheless, it is merely a philosophical truth. In order to actually realize or feel unity one must go beyond the intellect. In fact, even the mind must be suspended in order to experience the unity that exists in pure consciousness. For it is only through determined effort that we can obtain direct experience of unity and be fully ensconced in the ecstasy of oneness.

Glossary

ACHARYA: One who teaches by example. Spiritual teacher qualified to teach all lessons of meditation.
ADVAITA: Sanskrit term meaning non-duality or monism.
ADVAITA VEDANTA: The monistic philosophy of Shankara
AERIAL FACTOR: *Vayutattva*. One of the five fundamental factors created from Cosmic Mind-stuff (*Citta*). The primordial aerial factor was hydrogen, which condensed to form stars.
AHAMTATTVA, AHAM: Doer "I," ego, mental subjectivity, created from *Mahat* or "I feeling" by the action of the mutative principle or *rajoguna*.
AKASHA: Ether, space-time. The subtlest of the five fundamental factors.
ANNAMAYA KOSHA: The body or crudest layer of unit mind.
ANANDA: Divine bliss, ecstasy.
ANANDA MARGA: Path of divine bliss.
ANTHROPIC PRINCIPLE: The idea that without conscious beings the universe could not exist as it does.
ARCHETYPE: The idea, originally proposed by Plato that for every physical manifestation there is a mental manifestation. Hence, for every physical object and for every human idea there must be first a corresponding mental form or archetype existing in the Cosmic *Citta* (Mind-stuff). The first mental forces originating from Cosmic Consciousness.
ANTIMATTER: Matter with exactly the opposite characteristics of normal matter. When exposed to normal matter, antimatter is destroyed with the release of tremendous energy.
ASANAS: Postures of hatha yoga.
ASHTANGA YOGA: The eight-limbed path of yoga consisting of *yama* (do's), *niyama* (don'ts), asanas (postures), pranayama (breath control), *pratyahara* (withdrawal from sense and motor organs), *dharana* (concentration), dhyana (meditation), and samadhi (absorption in the Great).

ASTRAL: Pertaining to the unconscious (superconscious) layers of mind.
ASTRAL PROJECTION: When mind becomes detached from the body but is still connected to it via a thin thread. The mind in such a state is able to perceive events taking place at a remote location but is unable to affect physical reality in any way. See also, remote viewing.
ATMAN: Soul, unit consciousness, the atman of the cosmos is Paramatman, and that of the unit is the atman or *jivatman*.
ATIMANASA KOSHA: The first layer of the unconscious mind responsible for intuition, creativity, and ESP.
AVADHUTA: Literally, "one who is thoroughly cleansed mentally and spiritually"; a monk or nun who follows the tradition of Shiva Tantra.
AVATAR: God descending into human form. For example, Christians believe Jesus was an Avatar. Similar to *Mahasambhuti*.
AVIDYA: Ignorance. Retrograde movement of the Brahma Chakra. Movement from subtle to crude.
AYURVEDA: Ancient system of medicine born in India.
BABA: The Lord; Supreme Father; *Paramapurusha*.
BABA NAM KEVALAM: Sanskrit universal mantra meaning "everything is the One."
BELL'S THEOREM: A theorem by Bell proving that local hidden variables are incompatible with quantum mechanics since they would have to exist in a world outside space-time.
BHAGAVAN: One who has all the qualities of the Supreme Entity; the Lord.
BHAGAVAT DHARMA: The dharma to attain the Supreme. Human dharma.
BHAKTI: Devotion; love of God.
BHAKTI YOGA: The yoga of love or devotion. The highest form of yoga.
BIG BANG THEORY OF THE UNIVERSE: Theory that creation began from a singularity some fourteen billion years ago in a giant explosion.
BLACK HOLE: The densest form of matter known to exist in the universe. The gravitational force of a black hole is so large that it "ruptures" the fabric of space-time such that even light cannot escape its pull. Sometimes ambiguously called a singularity.
BLOCK TIME: Time as an unchanging fourth dimension with space. All past, present, and future events are already present at each point in four-dimensional space-time, but are not perceived by the unit mind since it has perception in only three-dimensional space and perceives time as the passage of events. The subtler Cosmic Mind, which is outside space-time,

can perceive block time and thus all events past, present, and future.
BRAHMA: Supreme Entity; Cosmic Entity. The One, which is both inside and outside everything, and therefore has no second (advaita). Brahma has complementary aspects of *Purusha* (Consciousness) and *Prakriti* (Qualifying Principle).
BRAHMA CHAKRA: Cycle of Creation. How Supreme Consciousness is transformed into the material world and into living creatures; has both an extroversive phase (*saincara*) in which Consciousness is transformed into crude (mind and matter) and an introversive phase (*prati-saincara*) in which matter is transformed into living organisms.
BUDDHA: The Enlightened One.
BUDDHI: Intellect.
CAUSAL DETERMINISM: One of the doctrines of materialism that cause always follows effect and that if the initial conditions of a system were known exactly then the exact path and velocity of an object could be predicted exactly. Has been conclusively demonstrated not to apply to quantum particles.
CHAKRA: Circle; psychic nerve plexus.
CITTA: Objective Mind-stuff; mental ectoplasm; done or objective "I."
COLLECTIVE UNCONSCIOUS: Term coined by Carl Jung to explain the aspect of the human mind that is beyond space, time, and person. Can be considered to be identical to the unconscious or superconscious mind consisting of *atimanasa, vijinanamaya, and hiranmaya koshas.*
COMPLEMENTARITY: A characteristic of quantum systems and of nature to exist in different, mutually exclusive aspects such as waves and particles. Both aspects are required in order to obtain a complete picture of the system.
CONSCIOUSNESS: The typical meaning of consciousness is a state of awareness. When capitalized in this book Consciousness refers to the subtle, unqualified stuff from which everything originates, and is both the beginning and endpoint of creation. Hence, this Supreme Consciousness is both the witness of creation and also the substance of creation and is also referred to as Cosmic Consciousness, Paramatman, and *Paramapurusha*. Unit consciousness or atman is associated with individual entities and is beyond time so it is neither created nor destroyed but eventually it will become one with Cosmic Consciousness.
COSMIC ENTITY: Brahma, God.
COSMIC MIND: The mind of God, consisting of *Mahattattva* (I feeling), *Ahamtattva* (doer I), and *Citta* (objective Mind-stuff).

DARK MATTER: The substantial quantity of invisible matter that is believed to exist because of its gravitational pull within galaxies. Most likely dark matter consists of more than a single unseen particle. Some may be molecular hydrogen (aerial factor) that has continued forming from ethereal factor since the Big Bang.
DARSHAN: Being in the physical presence of a spiritual master or guru. Literally, seeing the master.
DEATH: Irreversible separation of the mind and body. The bodiless mind will seek out a new physical structure so that it can continue on the path to final liberation.
DEVA: A god or deity. Philosophically, any vibration or expression emanating from the Cosmic Nucleus (*Purushottama*).
DEVAYONI: A luminous body; an advanced soul that performed sadhana but had a strong desire at the time of death causing it to take on a body consisting of ethereal, aerial, and luminous factors.
DEVOTION: See bhakti.
DHARMA: Characteristic property; innate spirituality; the path of righteousness.
DHARMACHAKRA: Group or collective meditation. Literally, circle of dharma.
DHARANA: Concentration of the mind. See also Ashtanga Yoga.
DHYANA: Meditation. Ideation on the Supreme Entity.
DUALISM: The philosophy that the ultimate reality consists of two or more separate things, such as mind and matter, or God and his creation.
ECTOPLASM: *Citta*, Mind-stuff.
ECSTASY: Ananda; samadhi, feeling that accompanies unification with the One.
EGO: Self-image; unit *Ahamtattva* or the portion of the unit mind that acts.
ELECTROMAGNETISM: One of the four fundamental forces of the physical world. Responsible for the attraction between oppositely charged particles, magnetism, and the chemistry and biochemistry of life.
ENLIGHTENMENT: Awakening; a heightened state of awareness in which the ego is inactive and a person feels that his every action and thought is the work of God.
ENTROPY: The tendency of the material world toward greater randomness or disorder.
EPIPHENOMENON: A secondary phenomenon dependent on something else. For example, material realism purports that mind is an epiphenomenon of brain.

ESP: Extrasensory perception; psychic abilities; parapsychology. Examples include clairvoyance, telepathy, remote viewing, precognition, and telekinesis.

EVIL: Movement from subtle to crude; action performed out of ignorance; *avidyamaya*.

FIVE FUNDAMENTAL FACTORS: Created from *Citta* by the action of the static binding force or *tamoguna*. They are in increasing crudity: ethereal factor (*akasha*), aerial factor (*vayutattva*), luminous factor (*tejastattva*), liquid factor (*apatattva*), and solid factor (*kshititattva*).

FREE WILL: The idea that human beings have the ability to determine how they act or direct their mind. Exists only in organisms that have a preponderance of *Mahattattva* or "I feeling."

GAUGE PARTICLES OR BOSONS: In the Standard Model of physics the gauge bosons are virtual particles that are exchanged in the creation of the fundamental forces. Photons for the electromagnetic force; W and Z bosons for the weak force; and gluons for the strong force.

GOOD: Movement from crude to subtle; action performed from knowledge; *vidyamaya*.

GRAND UNIFIED THEORY (GUT): A theory of physics that attempts to unify the electromagnetic, weak, and strong interactions into one unified model. So far all attempts to include gravity into the theory have met with failure.

GRAVITON: The theoretical force particle exchanged between particles possessing mass and responsible for the force of gravity. Has never been observed.

GUNA: Binding factor or principle; attribute; quality. *Prakriti*, the Cosmic Operative Principle, is composed of *sattvaguna*, the sentient principle; *rajoguna*, the mutative principle; and *tamoguna*, the static principle.

GURU: A spiritual teacher or prophet. Literally, one who dispels darkness.

HATHA YOGA: Yoga practice that emphasizes purification of the body and the endocrine glands; yoga postures or asanas.

HIDDEN VARIABLES: The theory postulated by Bohm and others that there exist unknown variables affecting quantum systems that are responsible for nonlocality and thus material realism is upheld. However, Bell's theorem proved that such variables would have to exist outside space-time and would effectively violate material realism.

HIRANMAYA KOSHA: The subtlest layer of unit mind. Cosmic *Mahattattva* is fully reflected in this layer of mind and when fully ensconced in this layer of mind the aspirant feels that they are one with God (*savikalpa* samadhi).

HOLISM: The philosophy that everything is a manifestation of the One; the Unity Principle.
IDEALISM: The philosophy that mind is a product of Consciousness and matter is a product of mind. See also Brahma Chakra.
IISHVARA: Brahma; the Cosmic Controller; literally, "the Controller of all controllers."
INDRIYA: One of the five sensory organs or five motor organs. For example, the eye *indriya* comprises the eye itself, the optic nerve, and the location in the brain at which the visual stimulus is transmitted to the ectoplasm, or mind-stuff.
ISHTA: One's individuality or personal goal.
ISHTA MANTRA: A personal mantra that is correlated with the breathing and is incantative, pulsative, and ideative (has meaning: usually a name for the Lord).
JADASPHOTA: Sanskrit term for a supernova
JAPA: The silent or mental repetition of a mantra.
JIVA: An individual being.
JIVATMAN: Unit consciousness.
JINANA: Knowledge; understanding.
JINANA YOGA: A type of yoga that emphasizes discrimination or intellectual understanding.
KARMA: Action; either positive or negative. An action that produces samskaras.
KARMA YOGA: A type of yoga practice that emphasizes selfless action.
KIRTAN: Collective singing of the name of the Lord, sometimes combined with a dance that expresses the spirit of surrender.
KOAN: An irrational question or statement utilized in Zen Buddhism to stimulate the mind to think "outside the box" of normal reality and induce an enlightened state of mind.
KUNDALINI: The coiled or latent spiritual energy residing in the *muladhara* or first chakra located at the base of the spine.
LIILA: Divine play. The creation is sometimes called the *liila* of the Supreme Entity.
LOKA: A "level" or "layer" of the Macrocosmic or Cosmic Mind.
LALITA: Spiritual dance designed to "open the heart," often practiced with kirtan.
LUMINOUS BODY: A living being with a body composed only of ethereal, aerial, and luminous factors; a *devayoni*. There are seven basic types

depending on their mental propensities.
MADHUVIDYA: Knowledgeable action; action performed without ego; action that does not create samskaras.
MAHABHARATA: "Great India;" the name of a military campaign guided by Lord Krishna around 1500 BCE to unify India; also the epic poem written by Maharshi Vyasa about this campaign.
MAHASAMBHUTI: When the Supreme Entity utilizes the five fundamental factors to express himself through a body, this is known as his *mahasambhuti*. Examples are Sadashiva, Krishna, and Anandamurti.
MAHATTATTVA, MAHAT: "I Feeling" ("I am," "I exist"); first subjectivity created from Consciousness and subtlest component of Cosmic Mind.
MANOMAYA KOSHA: The subconscious layer of mind that is responsible for higher thinking, contemplation, reasoning, pleasure, and pain. Philosophies, scientific theories, and all sorts of problem solving activity take place in this layer of mind.
MANTRA: A sound or collection of sounds (usually from the Sanskrit language) which, when meditated upon, will lead to spiritual liberation.
MARGA: Path.
MARGII: Someone on the path of bliss; a member of Ananda Marga.
MATERIALISM: Material realism; the philosophy that the material universe is the ultimate reality and that everything including mind and consciousness developed from matter and energy.
MAYA: Illusion. The Creative Principle or *Prakriti*. Also, the power of the Creative Principle to cause the illusion that finite created objects are the ultimate truth.
MEDITATION: Sadhana; any practice in which the mind is concentrated and the "I feeling" of an individual is subjectified.
MENTAL PLATE: The portion of the brain in which any form of awareness takes place.
MICROVITUM, MICROVITA: The smallest unit(s) of the life force.
MIND: For units (unit mind) the non-physical organizational function of an organism associated with the brain or nerve plexuses. Unit mind originates in the first phase of *prati-saincara* when under the right conditions pre-biotic chemicals arrange themselves under the intense pressure of Cosmic *Citta* to form simple living organisms. See also Cosmic Mind.
MOKSHA: Spiritual emancipation, non-qualified liberation. Merger with *Nirguna* Brahma upon physical death.
MONISM: The philosophy that there is only one entity making up the universe—the *Paramapurusha* (Supreme Consciousness) and that

everything is derived from it.
MONISTIC IDEALISM: See the Unity Principle.
MUDRAS: Spiritually symbolic hand or body gestures or movements.
MUKTI: Spiritual liberation. Merger with *Saguna* Brahma following physical death.
MYSTIC: Someone who taps into the limitless, timeless, egoless unconscious mind.
MYSTICAL EXPERIENCE: The experience of a mystic; an ego-loss experience; feeling of being one with God.
NADI: Psychic nerve or channel.
NARAYANA: Another name for the Supreme Entity; literally, "the Lord of *Nara*" (*Prakriti*).
NIRGUNA BRAHMA: Brahma unaffected by the *gunas* or *Prakriti*; non-qualified Brahma.
NONLOCALITY: In quantum mechanics, it is the instantaneous influence or communication of information through space-time. Philosophically the connectedness permeates everything because ultimately everything is a manifestation of the One.
OM: The sound of the first physical (as opposed to mental) vibration of creation; the sound emitted by *akasha* (space-time).
OPERATIVE PRINCIPLE: Another name for *Prakriti*.
PANENTHEISM: The philosophy that the creation is God and that he also transcends it.
PANTHEISM: The philosophy that God is the creation and the creation is God.
PAPA: Sin; action performed out of ignorance (*avidya*).
PARAMAPURUSHA: Supreme Consciousness; Cosmic Consciousness.
PARAMATMAN: Supreme Consciousness, Cosmic Consciousness, *Purushottama* in the role of the witness of creation.
PHOTON: The particle aspect of light. The particle exchanged between charged matter particles giving rise to the electromagnetic force.
PRAKRITI: Cosmic Operative Principle. Composed of the three *gunas* that qualify or transform Consciousness.
PRANA: Life force; vital energy; psychic energy. The energy that runs the psychic body and its chakras. Called chi in Chinese.
PRANAYAMA: Breath control; one of the practices of Ashtanga Yoga.
PRATI-SAINCARA: In the Cosmic Cycle of Creation (Brahma Chakra), the systematic introversion and subtilizing of Consciousness from the state of solid matter back to the Nucleus Point of creation (*Purushottama*). *Prati*

means "counter" and *saincara* means "movement."
PRATYAHARA: Withdrawal of the mind from the sense and motor organs.
PURUSHA: Consciousness. The witnessing aspect of Brahma, the Cosmic Entity. *Paramapurusha* is Supreme or Cosmic Consciousness. An individual's unit consciousness is normally called atman.
PURUSHOTTAMA: The Nucleus of all creation, the witness of qualified creation. The *Saguna* Brahma.
QUANTUM: The lowest denomination of matter or energy that can be exchanged. It is a discreet packet of energy that transitions between energy levels by jumping without going through intermediate levels; e.g. a photon.
QUANTUM MECHANICS: Quantum theory; quantum physics; the branch of physics that provides a mathematical description for the behavior and interaction of matter and energy on the sub-atomic (quantum) scale.
RAJOGUNA: The mutative binding force of *Prakriti*. When it acts upon *Mahattattva* the *Ahamtattva* is created.
REINCARNATION: The doctrine that the unit consciousness (atman) along with its associated samskaras survives physical death and with the help of the Cosmic Mind eventually finds a suitable physical body to continue its journey to ultimate perfection or merger with Supreme Consciousness.
REMOTE VIEWING: A form of ESP in which someone is able to perceive images or events taking place at a location distant from their body. Similar to astral projection.
RISHI: A sage or guru; one who is instrumental in advancing human civilization along the path toward enlightenment.
SADASHIVA, SHIVA: A great Tantric guru of 5000 BCE who was instrumental in advancing the science of Tantra. Considered a *mahasambhuti* or a prophet who was fully liberated while living in a human body.
SADHAKA: Spiritual aspirant or practitioner.
SADHANA: Literally, "sustained effort"; spiritual practice; meditation.
SAGUNA BRAHMA: Brahma affected by the *gunas*; qualified Brahma.
SAHASRARA CHAKRA: The highest chakra located at the crown of the head. The thousand-petaled lotus.
SAINCARA: In Brahma Chakra (the Cosmic Cycle), the step-by-step extroversive movement and condensation of Consciousness from the Nucleus Point of creation (*Purushottama*) to the state of solid matter (*saincara* literally means "movement").
SAMADHI: Merger of the unit mind into the Cosmic Nucleus (*savikalpa* samadhi) or into unqualified Supreme Consciousness (*nirvikalpa* samadhi). Samadhi occurs while in the body. Upon death *savikalpa* samadhi

corresponds to mukti and *nirvikalpa* samadhi corresponds to moksha.
SAMBHUTI: The whole of creation, see also *mahasambhuti*.
SAMSKARA: Mental reactive momentum, potential reaction to some action or thought performed previously.
SANSKRIT: The ancient language of the Vedas, composed of fifty sounds that correlate to the fifty *vrittis* or propensities of the human mind. The only language known with the subtlety to express the ideas of advaita cosmology (the cosmology of the One or Unity Principle). Mantras are normally from Sanskrit.
SATTVAGUNA: The subtlest binding force of *Prakriti*, which creates the "I am" or *Mahattattva* from pure Consciousness.
SATYA: Truth; absolute reality.
SATSANG: Company of spiritually minded people; good company.
SELF: The subjective component of mind; the unit *Mahattattva* or "I feeling."
SELF-CONSCIOUSNESS: Awareness of one's own being as separate from the surroundings and other beings. Only humans are thought to have fully developed self-consciousness and therefore must take responsibility for their actions.
SENTIENT FORCE: *Sattvaguna*. The subtlest binding force of *Prakriti* that creates the subjective *Mahattattva* or "I exist feeling" from pure Consciousness.
SEVA: Service.
SHAIVITES: Worshippers of Shiva.
SHAKTI: Power; also *Prakriti*.
SHASTRA: Scripture.
SHIVA: See Sadashiva.
SHLOKA: A Sanskrit aphorism expressing a single idea.
SCHRÖDINGER WAVE FUNCTION: The mathematical description of a quantum field such as an electron. When squared it can give a probability distribution for the particle.
SHUNYA: The Buddhist doctrine of the Void.
SIDDHA: A spiritually advanced soul; Hindu ascetic who has attained enlightenment or become a *devayoni*.
SIN: Action performed out of ignorance (*avidya*). Also *avidyamaya*.
SPACE-TIME: Ether; *akasha*; the four-dimensional nature of space entwined with time. See also block time.
SPIRITUALITY: The concept that there exists an ultimate reality that transcends the material world; the concept that mind and matter are

derived from spirit (i.e. Consciousness). The belief that we have a soul or spirit that survives death and that union with Cosmic Consciousness is the ultimate purpose of human existence.
SOUL: Unit consciousness or atman.
STATIC FORCE: See also *tamoguna;* the strongest binding force or *guna* of *Prakriti.*
STRONG OBJECTIVITY: One of the postulates of material realism; the idea that objects are separate and no way connected and are independent of the observer. Does not hold for quantum objects and systems.
SUPERCONSCIOUS MIND: Unconscious mind. The subtle, nonlocal layers of individual mind consisting of *atimanasa, vijinanamaya,* and *hiranmaya koshas.* Sometimes referred to as causal mind.
SUPERNOVA: Also *jadasphota*; the explosion of a massive star that occurs when it expends its thermonuclear fuel and collapses resulting in a tremendous explosion and release of energy whereupon heavy elements beyond iron are formed. New stars are often born from the gas and particle remnants of a supernova.
TAMOGUNA: The static binding force of *Prakriti.* Acting upon *Ahamtattva* it creates *Citta* and when *Citta* is bound further by *tamoguna* the five fundamental factors are created.
TANDAVA: Dance introduced by Sadashiva to overcome fear.
TANMATRAS: Inferential waves that affect the sense organs. The vibrations produced by the five fundamental factors that can be sensed.
TANTRA: A spiritual tradition that originated in India in prehistoric times and was first systematized by Sadashiva. It is designed to enhance personal and spiritual energy and overcome all fears and weaknesses through meditation and exposure to the difficulties inherent in the world.
TARAKA BRAHMA: The Supreme Entity in its liberating aspect. It is the *Taraka* Brahma that takes human form as his *mahasambhuti* in order to help liberate humankind from crudity.
TATTVA: Factor, such as *Ahamtattva,* the mental factor of acting or *tejastattva* the luminous factor.
TEJASTATTVA: Luminous factor; electromagnetic radiation.
TELEOLOGY: Goal-directed activity; the concept that there is a final goal or endpoint to creation.
TELEOLOGICAL EVOLUTION: The theory that evolution is not merely chaotic but has an underlying purpose and endpoint since it is guided by the Cosmic Entity.
THERMAL DEATH OF THE UNIVERSE: The hypothesis that because

entropy increases in nearly all reactions of matter the universe is tending toward a state of increasing randomness and thermal consistency, whereupon eventually it will attain a uniform temperature and cease to produce any new stars or galaxies. Could occur if all the matter and energy of the universe was created in the Big Bang but is negated by the fact that new matter (hydrogen) is continuously being created from the ether.

TRANSCENDENTAL ENTITY: Brahma; *Paramapurusha*, God, etc.

UNCERTAINTY PRINCIPLE: The principle first postulated by Heisenberg that the complementary aspects of a quantum object such as momentum and position cannot be measured with absolute accuracy.

UNIT CONSCIOUSNESS: Atman or soul. Unit Mind: see Mind.

UNITY PRINCIPLE: Monism; monistic idealism; holism; monistic panentheism; non-dualism, inseparability, fundamental oneness, advaita. The concept that ultimately the created universe is a Singularity or One that is created from Supreme Consciousness by the action of the Qualifying Principle, *Prakriti*. See also Brahma Chakra.

UPANISHADS: Vedic scriptures that form the core teachings of Hinduism, i.e., the monism of the Unity Principle.

VEDANTA: Ancient eastern philosophical tradition concerned with understanding the ultimate reality (Brahma).

VEDAS: Ancient verses originating in India and later written in Sanskrit. The texts constitute the oldest scriptures of Hinduism.

VIJINANAMAYA KOSHA: The second layer of the unconscious or superconscious mind responsible for divine attributes such as spiritual ecstasy, humility, meditation, magnanimity, discrimination, and non-attachment.

VIDYA: Knowledge; the path of liberation; movement from crude to subtle.

VITAL ENERGY: Prana; psychic energy that animates the psychic body and chakras.

VRITTIS: Propensities of mind; desires.

WAVE FUNCTION: Quantum probability waves resulting from the solution of the Schrödinger equation. Used to mathematically describe the quantum state of a particle or system. Collapse of the wave function into a specific state occurs upon measurement or observation.

YOGA: Literally, to yoke or unite. Spiritual practice leading to unification of the unit soul or atman with the Supreme Consciousness (Paramatman).

Notes

Introduction

1. Dean I. Radin, *The Conscious Universe: The Scientific Truth of Psychic Phenomena* (New York: HarpersEdge, 1997).
2. Raymond A. Moody, Jr. and Paul Perry, *Coming Back: a Psychiatrist Explores Past-life Journeys* (New York: Bantam, 1991).
3. Bernard d'Espagnat, *In Search of Reality* (New York: Springer-Verlag, 1983).

Chapter 1. The New Physics of Quanta

1. *Science News*, July 22, 1997.
2. Menas Kafatos, and Robert Nadeau, *The Conscious Universe: Part and Whole in Modern Physical Theory* (New York: Springer-Verlag, 1990), 165. The authors point out that space and time nonlocalities inexorably point to an undivided wholeness in the cosmos. Quantum fields are nonlocal, meaning the field is not located in a given region of space and time. In a nonlocal phenomenon, what happens in region A instantaneously influences what occurs in region B, and vice versa, without the exchange of energy. Nonlocality has been dramatically and convincingly revealed by experiments in modern physics and the observations are independent of any theory, which means that any future theory of nature must also embody nonlocality.

Chapter 2. Time, Space, and Relativity

1. Stephen Hawking and Roger Penrose, *The Nature of Space and Time* (Princeton: Princeton University Press, 1996), 89-90.

2. Eckhart Tolle, *The Power of Now: a Guide to Spiritual Enlightenment.* (Vancouver: Namaste Publishing, 1999).
3. The undetected dark matter has been shown to be slowing the rotation of galaxies and its gravitational pull bends the light passing near distant galaxies. Measurements of the cosmic microwave background radiation or afterglow of the Big Bang show that the early universe consisted of 15 percent ordinary matter and 85 percent dark matter. Dark matter may consist of more than one particle. Some dark matter may be molecular hydrogen (H_2). Molecular hydrogen, unlike atomic hydrogen (H), is invisible except at certain infrared wavelengths. Recent studies using the European Space Agency's Infrared Space Observatory indicate that molecular hydrogen in galaxies may exceed atomic hydrogen by a factor of ten to one, which would account for much of the missing mass in the universe. Since atomic hydrogen is less stable than molecular hydrogen, over time most of the hydrogen in the universe may have combined to make the molecular form.

Chapter 3. Simple is Better

1. Paul Davies, *In Who We Live and Move and Have Our Being: Panentheistic Reflections on God's Presence in a Scientific World.* Editors: Philip Clayton and Arthur Peacocke (Grand Rapids, MI: Eerdmans Publishing, 2004).
2. Dean I. Radin, *Entangled Minds, Extrasensory Experiences in a Quantum Reality* (New York, NY: Paraview Pocket Books, 2006).
3. Maryann Mott, *National Geographic News.* January 4, 2005.
4. Don Oldenburg, *Washington Post.* January 8, 2005.
5. Maryann Mott, *National Geographic News.* November 11, 2003.
6. Stephen Hawking and Roger Penrose, *The Nature of Space and Time* (Princeton: Princeton University Press,1996), 89-90.
7. Roger Penrose, *The Road to Reality: A Complete Guide to the Laws of the Universe* (New York: Knopf, 2005), 726-756.
8. Martin J. Rees, *Just Six Numbers: The Deep Forces That Shape the Universe* (New York: Basic Books, 2000).
9. Garret Lisi, "An exceptionally Simple Theory of Everything,*" J. of High Energy Physics,* 711, Nov. 2007.
10. Michael J. Behe, *Darwin's Black Box: The Biochemical Challenge to Evolution* (New York: Free Press, 1996).

Chapter 4. Brahma Chakra: The Cycle of Creation

1. William A. Tiller., Walter Dibble, and Michael Kohane, *Conscious Acts of Creation: The Emergence of a New Physics* (Walnut Creek, CA: Pavior Publishing, 2001).
2. For example, many new stars can be observed forming in the Crab Nebula in the constellation Orion from the remnants of a supernova that was observed there by Arab astronomers in 1054 AD.
3. Rudreshananda, *Microvita: Cosmic Seeds of Life* (Mainz, Germany: DharmaVerlag, 1989).
4. Michael J. Behe, *Darwin's Black Box: The Biochemical Challenge to Evolution*.
5. For a comprehensible discussion of this problem with material Darwinism see: Thomas Nagel, *Mind and Cosmos: Why the Materialist neo-Darwinian Conception of nature is Almost Certainly False* (Oxford: Oxford University Press, 2012).
6. Larry Dossey, *Recovering the Soul: a Scientific and Spiritual Search* (New York: Bantam, 1989), 113.

Chapter 5. The Rule of Three

1. The graviton is more a theoretical particle since it has never been actually observed.
2. Leon Lederman and Dick Teresi, *The God Particle: If the Universe Is the Answer, What Is the Question?* (New York: Houghton Mifflin, 1993).
3. Deepak Chopra, *The Way of the Wizard* (New York: Harmony Books, 1995), 161.
4. Carl G. Jung, *Modern Man in Search of a Soul* (New York: Harcourt Brace Jovanovich, 1933).

Chapter 7. The Nature of Unit Consciousness and Unit Mind

1. Clearly mind does affect matter since the simple act of moving a finger involves a mental action that affects a material object (a nerve). However, this interaction takes place at the quantum level in the uncertainty gap defined by Planck's constant when the wave function is collapsed. This interaction of mind and matter cannot be observed by any measuring device since measurement itself affects what is measured at the quantum level.
2. Raymond A. Moody Jr., *Coming Back: a Psychiatrist Explores Past-life Journeys*.

Chapter 9. Life after Death

1. Raymond A. Moody, Jr., *Coming Back: a Psychiatrist Explores Past-life Journeys.*
2. Brian L. Weiss, *Many Lives, Many Masters* (New York: Simon & Schuster, 1988).
3. Raymond A. Moody, Jr., *Life after Life: The Investigation of a Phenomenon—Survival of Bodily Death* (New York: HarperCollins, 1975).
4. Jeffrey Long and Paul Perry, *Evidence of the Afterlife: The Science of Near-Death Experiences* (New York: HarperCollins, 2010).
5. Larry Dossey, *Recovering the Soul: a Scientific and Spiritual Search*, 17-19.
6. Elaine Pagels, *The Gnostic Gospels* (New York: Random House, 1979).
7. Matthew 11: 13-14; Matthew 17: 10-13.
8. Matthew 16:3; Luke 9:18; Mark 8:26.
9. Herbert B. Puryear, *Why Jesus Taught Reincarnation* (Scottsdale, AZ: New Paradigm Press, 1993).
10. Quincy Howe, Jr., *Reincarnation for the Christian* (Philadelphia: Westminster Press, 1974).

Chapter 10. Psychic Body and Power

1. For a discussion of dark matter see Chapter 2.
2. Dean I. Radin, *The Conscious Universe: The Scientific Truth of Psychic Phenomena* and *Entangled Minds, Extrasensory Experiences in a Quantum Reality.*
3. William A. Tiller, Walter E. Dibble, and Michael J. Kohane, *Conscious Acts of Creation: The Emergence of a New Physics.*

Chapter 11. Meditation

1. Anandamurti, *Namah Shivaya Shantaya* (Purulia, India: Ananda Marga Publications, 1982), 8.

Chapter 12. The Unity Principle in the Teachings of Religious Prophets

1. Anandamurti, *Namah Shivaya Shantaya.*
2. Elaine Pagels, *The Gnostic Gospels.*

3. Elaine Pagels, *Beyond Belief: The Secret Gospel of Thomas* (New York: Random House, 2003).
4. Deepak Chopra, *The Third Jesus* (New York: Harmony Books, 2008), 161.

Chapter 13. The Unity Principle in Scripture

1. Aldous Huxley, *The Perennial Philosophy* (New York: Harper Perennial, 1944), 7-8.
2. Anandamurti, *Ananda Sutram* (Kolkata: Ananda Marga Publications, 1962).
3. Ananda Mitra, *The Spiritual Philosophy of Shrii Shrii Anandamurti, a Commentary on Ananda Sutram* (Kolkata: Ananda Marga Publications, 1981).

Chapter 14. Scientists on Unity

1. Albert Einstein quoted in the *New York Post*, November 28, 1972.
2. Albert. Einstein, *Ideas and Opinions* (New York: Three Rivers Press, 1954), 12.
3. Max Planck quoted in *Accent Magazine*, "The Hidden World of Mind," Oct. 1972, New Delhi, India.
4. Wolfgang Pauli, *Writing on Physics and Philosophy* (Berlin: Springer-Verlag, 1994), 259.
5. Ken Wilber, *Quantum Questions, Mystical Writings of the World's Greatest Physicists* (Boston: Shambhala, 1984), 173
6. Ibid., 175
7. Ibid., 144
8. Ibid., 150
9. Ibid., 145
10. Erwin Schrödinger, *What is Life* (Cambridge: Cambridge University Press, 1947), 137.
11. Erwin Schrödinger, *My View of the World* (Woodbridge, CT: Ox Bow Press, 1983), 31-4.
12. Ken Wilber, *Quantum Questions, Mystical Writings of the World's Greatest Physicists*, 87.
13. Werner Heisenberg, *Across Frontiers* (Woodbridge, CT: Ox Bow Press, 1990), 105-6.
14. Ibid., 116-7.

15. Werner Heisenberg, *Physics and Philosophy: The Revolution in Modern Science* (New York: Harper & Row, 1962), 186.
16. Ibid., 71
17. Ibid., 191-2.
18. David Bohm, *Wholeness and the Implicate Order* (New York: Routledge & Kegan Paul, 1983), 175.
19. Ibid., 7.
20. Ibid., 191-2.
21. Ibid., 210.
22. Ibid., 16.
23. Larry Dossey, *Recovering the Soul: a Scientific and Spiritual Search*, 175.
24. Henry Margenau, *The Miracle of Existence* (Woodbridge, CT: Ox Bow Press, 1984), 120.
25. Ibid., 96.
26. Fritjof Capra, *The Tao of Physics* (Boulder, CO: Shambhala Publications, 1975), 116-117.
27. Ibid., 117-118.
28. Ibid., 69.
29. Menas Kafatos, and Robert Nadeau, *The Conscious Universe: Part and Whole in Modern Physical Theory*.
30. Amit Goswami, *The Self-Aware Universe: How Consciousness Creates the Material World* (New York: Tarcher/Putnam, 1993).
31. For example, Deepak Chopra (*Quantum Healing* plus numerous other titles), Larry Dossey (*Recovering the Soul*), William Tiller, Walter Dibble, Michael Kohan (*Conscious Acts of Creation*), Ken Wilber (*Quantum Questions, The Holographic Paradigm*), Fred Wolf (*The Yoga of Time Travel*), Arthur Zajonc (*Catching the Light*), Eckhart Tolle (*The Power of Now, A New Earth*), Ram Dass (*Being There, Still Here, Paths to God*), Jeffry Satinover (*The Quantum Brain*), Danah Zohar (*Quantum Self, Quantum Society*), Andrew Newberg (*Why God Won't Go Away: Brain Science & the Biology of Belief*), William Arntz, Betsy Chasse, Mark Vicente (*What the Bleep Do We Know*), Gary Zukav (*Dancing Wu Li Masters: An Overview of the New Physics*), Michael Towsey (*Eternal Dance of Macrocosm*).

Chapter 15. Western Attempts to Understand Unity

1. For an excellent review on western panentheism see: Philip Clayton and Arthur Peacocke, Editors *In Whom We Live and Move and Have Our Being: Panentheistic Reflections on God's Presence in a Scientific World* (Grand

Rapids, MI: Eerdmans Publishing, 2004).

Chapter 16. The Mystical Vision of Unity

1. Robert Ullman and Judyth Reichenberg-Ullman, *Mystics, Masters, Saints, and Sages, Stories of Enlightenment* (Berkeley, CA: Conari Press, 2001).
2. Richard Bucke, *Cosmic Consciousness* (New York: Penguin Books, 1901).
3. Robert Ullman and Judyth Reichenberg-Ullman, *Mystics, Masters, Saints, and Sages*, 43.
4. Ibid., 43.
5. Ibid., 97.
6. Ibid., 193.
7. Gopi Krishna, *Living with Kundalini* (Boston: Shambhala, 1993), 2.
8. Eckhart Tolle, *The Power of Now: a Guide to Spiritual Enlightenment*, 12.
9. Timothy Leary, Ralph Metzner, and Richard Alpert, *The Psychedelic Experience: A Manual Based on The Tibetan Book of the Dead* (New York: Citadel Press, 1964).

Chapter 18. Practicing Unity

1. The unit consciousness or atman is simply a reflection of the Supreme Consciousness or Paramatman on the unit mental plate. The unit's sense of "I am" is nothing but Cosmic *Mahattattva* reflected on the mental plate. This "I" is universal and is witnessed by the atman. Therefore, when the personal ego is transcended all that remains is the transcendental experience of "I am that Absolute Truth."
2. Victor J. Stenger, *The Unconscious Quantum: Metaphysics in Modern Physics and Cosmology* (Amherst, MA: Prometheus 1995), 28.
3. We have already seen that science does not support the materialist-atheist-skeptic's view of the universe, nor could it ever answer the question of whom or what came before or initiated the Big Bang. Nonetheless, materialists believe that matter/energy is real and created consciousness. The material universe is the only reality and runs like a machine according to physical laws. Belief in God is a myth and science can eventually answer all the questions, and there is no need to postulate a Supreme Entity. Although science has answered many questions about the origin of the universe and development of living organisms, to believe that it ultimately

will provide all the answers is nothing but a faith-based doctrine similar to a belief in God taught by most religions.

4. Shrii Shrii Anandamurti, *Ananda Vanii Samgraha*, 2nd Edition, (Kolkata: Ananda Marga Publications), 21.
5. Ibid., 43-44.
6. Luke 18:17.
7. Mark 10:24, Matthew 5:44.
8. Matthew 5:39.
9. See for example Eckhart Tolle, *The Power of Now* and *A New Earth* and *Stillness Speaks*.
10. Deepak Chopra, *The Seven Spiritual Laws of Success* (San Rafael, CA: Amber-Allen, 1994).
11. Deepak Chopra, *The Way of the Wizard* (New York: Harmony Books, 1995).
12. For example, see Gary Crowley, *From Here to Here: Turning toward Enlightenment* (Boulder, CO: GL Design, 2006).
13. Ram Dass, *Paths to God: Living the Bhagavad Gita* (New York: Harmony 2004).

Index

A

acharya, 76, 130, 187
Adams, Robert, 157
advaita, 187, 189, 196, 198
Advaita Vedanta, 128, 187
aerial factor, 34, 54, 55, 57, 64, 187, 190, 191
Aham, 61, 62, 66, 76, 80, 81, 85, 90, 92, 93, 187
Ahamtattva, 53, 54, 59, 61, 63, 68, 84, 115, 118, 166, 173, 187, 189, 190, 195, 197
ajina chakra, 107, 108
akasha, (see also ether), 54, 63, 187, 191, 194, 196, 218, 221
Alpert, Richard, 159
anahata chakra, 107-109
ananda, 66, 92, 93, 124, 187, 190
Ananda Marga, 128, 187, 193, 202, 203, 206
Ananda Sutram, 134, 203
Anandamurti, 115, 120, 128, 129, 134, 154, 175-177, 202, 203, 206
animal instincts, 42, 43, 50, 170
annamaya kosha, 85, 88
anthropic principle, 45, 46, 47, 163
Anthroposophy, 130
antimatter, 33, 45, 187
Aquinas, St. Thomas, 150
archetype, 50, 137, 187
Arjuna, 121, 132, 133
asanas, 108, 125, 159, 180
Ashtanga Yoga, 125, 187, 219, 221
Aspect, Alan, 13, 14, 15, 147
astral, 2, 3, 188, 195
astral projection, 3
atimanasa kosha, 73, 87, 88, 108, 188
atman, 73, 85, 88, 89, 116, 124, 128, 157
avadhuta, 115, 128, 130, 188, 219
avatar, 121, 188
avidya, 65, 94-96, 113, 175, 188, 196
Ayurveda, 120, 188

B

Baba, 186, 188
Baba Nam Kevalam, 188
background radiation, 55, 70
Baha'u'llah, 51, 154
Behe, Michael, 49, 200, 201
Bell's theorem, 13, 163
Bhagavad Gita, 121, 132, 153

Bhagavan, 52
Bhagavat dharma, 188
bhakti, 179, 185, 186
Bhaktivedanta, A.C., 152
Big Bang, 16, 17, 18, 19, 26, 30, 34, 35, 45, 46, 47, 49, 52, 54, 55, 70, 106, 151, 163, 188, 205
Bird-of-Paradise, 43
black hole, 23, 31, 70, 168, 188
block time, 26, 29, 35, 151, 167, 169, 188, 189
blood clotting, 49, 61
bodiless mind, 83, 94, 100, 103, 104, 105, 110, 169
Bohm, David, 143, 146
Bohr, Neils, 12, 13, 135, 136
Brahma, 51, 52, 53, 54, 60, 63, 65-67, 78, 92, 95, 120, 128, 131, 144, 151, 152, 171, 183
Brahma Chakra, 51, 54, 60, 63, 65-67, 92, 95, 188, 189, 192, 194, 195, 198
Buddha, 51, 61, 83, 97, 112, 120, 123, 124, 154
buddhi, 61, 189
Buddhism, 5, 50, 51, 125, 127, 182

C

Campbell, Joseph, 50
Capra, Fritjof, 146, 147
Cayce, Edgar, 41, 111
causal determinism, 1, 4, 162, 189
chakra, 107, 108, 109, 117, 118, 125, 129, 192, 195
Chopra, Deepak, 76, 127, 130, 181, 201, 203, 204, 206
Christianity, 97, 98, 102, 103, 126, 149, 165

Chuang Tzu, 133
Citta, 53, 54, 56, 59, 61-64, 79, 83, 84-86, 88, 90, 100, 104, 105, 118, 166, 187, 189, 193, 197
clairvoyance, 40, 110, 191
clockwork universe, 22
collapse of the wave function, 9, 111, 145
collective unconscious, 50, 77, 88, 140, 187
complementarity, 7, 8, 137, 189
Confucius, 122
Consciousness, 3, 5, 6, 16, 26, 32, 33, 36, 38, 39, 47, 49, 51-54, 57, 65-68, 72, 77, 78, 81, 82, 84-86, 90, 92, 94, 95, 97, 99, 105-107, 108, 112, 114-116, 118, 124, 125, 134, 147, 152-155, 161, 166, 167, 169, 170, 174, 176, 179, 183-185
Constantine, 103, 126
Cosmic Entity, 33-36, 51-53, 58, 83, 108, 110, 114, 118, 120, 126, 131, 133, 139, 147, 148, 154, 155, 169-171, 173, 178, 179, 182, 183, 186, 189, 195, 197
Cosmic Mind, 34, 41-44, 49, 53, 54, 57, 59, 61-63, 65, 66, 68, 72-74, 77, 81, 83, 85, 87, 88, 90, 94, 95, 100, 104, 107, 108, 112, 134, 144, 167, 168, 169, 170, 172
Cosmic microwave background radiation, 55, 200
Cosmic Operative Principle, 194

D

d'Espagnat, Bernard, 5, 209
Dalai Lama, 130
dark energy, 19, 106
dark matter, 19, 35, 106, 190, 200,

202
darshan, 181
Darwin, 1, 60
Darwinism, 1, 151, 163, 169
Davies, Paul, 40
death, 2, 3, 13, 34, 35, 41, 44, 61, 62, 83, 94, 99, 100-103, 105, 110, 120, 121-124, 128, 129, 131-133, 140, 152, 154-156, 164, 169, 173
Descartes, Rene, 16, 139
deva, 190
devayoni, 105, 190, 192, 196
devotion, 105, 121, 127, 180, 185, 186
dharma, 96, 115, 121, 123, 133, 134
Dharmachakra, 190
dhyana, 118, 125
diffraction pattern, 11, 54
DNA, 39, 42, 43, 49, 61, 99, 163, 164
dual slit experiment, 9-11
dualism, 5, 98, 128, 152

E

Earth, 3, 17, 22, 24, 26-33, 42, 43, 45, 46, 54, 57-59, 63, 70, 71, 74, 156, 158, 165, 169, 177
ecstasy, 74, 88, 92, 154, 155, 156, 184, 186
ectoplasm, 189, 190, 192
Eddy, Mary Baker, 130
ego, 33, 51, 63, 65, 66, 74-77, 80, 81, 84, 88, 90, 92, 93, 102, 112-115, 117, 118, 121, 136, 145, 155, 159, 160, 171-179, 181-185
Eight-Fold Path, 124
Einstein, 7, 12, 13, 17, 22, 23, 27, 47, 54, 55, 106, 112, 135, 136, 167
electromagnetic force, 45, 68
electromagnetism, 47, 190

electroweak force, 44, 69, 70, 168
enlightenment, 3, 51, 92, 115, 124, 154, 161, 181, 182, 185
entropy, 19, 30, 190, 197
epiphenomena, epiphenomenon, 36, 164, 165, 36, 166
ESP, 2, 3, 110, 111, 164, 168
ether, 54, 167, 169
ethereal factor, 54, 55, 106
evil, 95-98, 191
evolution of species, 60, 169
Exceptionally Simple Theory of Everything, 47

F

fine-tuning, 19, 21, 35, 45, 46, 49, 71, 163
five fundamental factors, 57, 68, 85, 103, 105, 108, 120, 166, 167, 168, 187, 191, 197
Four Noble Truths, 124
free will, 26, 63, 65, 80, 151

G

Ganzfeld telepathy, 40
gauge bosons, gauge particles 69, 191
general theory of relativity (GTR), 23, 168
ghosts, 103-105
gilgul, 103
Gisin, Nicolas, 15
gluon, 69
gnostic Christians, 103, 127
gnostic Gospels, 125, 126, 202, 203
good and evil, 94, 95, 97, 151
Gopi Krishna, 154, 158, 205
Gospels, 103, 125, 126, 202

Grand Unified theory, 47, 191
gravitational lensing, 168
graviton, 47, 48, 69
gravity, 18, 19, 23, 32, 36, 44, 45, 47, 48, 69, 70, 71, 135, 163, 167, 168
guna, 52, 53, 78, 85, 191, 197
guru, 77, 112, 117, 118, 120, 181

H

Hadron Collider, 47, 70
happiness, 33, 66, 75, 76, 90, 91, 92, 93, 94, 96, 97, 124, 158, 179, 183,
Hasidic Judaism, 51, 103
Hatha yoga, 125, 180, 187, 191
Hawking, Stephen, 21, 45, 199, 200
Hegel, Georg, 151
Heisenberg, Werner, 8, 142, 198, 203
hidden variables, 163, 188, 191
Higgs field, Higgs boson, 51, 71
Hinduism, 50, 51, 52, 128, 131
hiranmaya kosha, 73, 87, 88, 108, 189, 191, 197
holism, 5, 15, 16, 34, 36, 51, 165, 175
Hubble, 17, 19
Huxley, Aldus, 119, 130, 160

I

idealism, 5, 147, 192, 198
Iishvara, 192
imprinted electrical devices, 14
indriya, 192
irreducibly complex, 49
ishta, 117, 118, 192
ishta mantra, 117, 192
Islam, 97, 127, 149, 165
Ismaili Islam, 127

J

jadasphota, 192, 197
japa, 117, 180, 181, 192
Jeans, James, 138, 139
Jesus, 103, 125, 126, 127, 154, 178, 181
jivas, 153, 192
jinana, 192
jinana yoga, 179, 192
Judaism, 51, 97, 102, 103, 121, 125, 149, 165
Jung, Carl, 50, 76, 88, 137, 140, 144

K

Kabbalah, 103
karma (see also Law of Karma), 41, 79, 80, 93, 180, 192
karma yoga, 180, 192
kirtan, 117, 180, 192
koan, 182, 192
kosha, 73, 85-88, 108, 109
Krishna, 120, 121, 129, 132, 133, 152, 153, 154, 158, 182
kundalini, 108, 117, 118, 129, 134, 158, 180, 192

L

lalita, 192
Lao Tzu, 51, 122, 154
Law of Karma, 41, 79, 80, 93
Lederman, Leon, 70, 201madhuvidya
Lie-group, 47
light, 1, 8, 9, 10, 12, 13, 14, 17, 18, 20-25, 27-34, 36, 45, 54, 55, 64, 69, 70, 95, 104, 106, 107, 135, 157, 158, 160, 167, 168
liila, 67

Lisi, Garret, 47
loka, 192
luminous body, 105, 190, 192
luminous factor, 55, 57, 64, 105, 168

M

madhuvidya, 81, 82, 178, 184, 186, 193, 212
Mahabharata, 121, 132
mahasambhuti, 120, 121, 129, 193, 195, 196, 197
Mahat, 85, 90, 92, 93, 178, 193, 187
Mahattattva, 53, 54, 59, 63, 64, 65, 81, 84, 85, 114, 115, 118, 154, 166, 169, 178, 189, 191, 193, 195, 196, 205
manipura chakra, 107-109
manomaya kosha, 86, 88, 108, 193
mantra, 117, 180, 181, 188, 192, 193
many-worlds, 40, 47, 163
marga, 66, 92
Margenau, Henry, 144-146
margii, 193
material realism, 1, 2, 4, 5, 12, 34, 39, 99, 174, 190, 191, 193, 197
materialism, 1, 3, 165, 169, 189, 193
maya, 87, 121, 152, 182, 193
meditation, 41, 65, 73, 75, 77, 78, 88, 92, 101, 108, 114-118, 124, 125, 127, 128-130, 159, 160, 176, 178, 180-182, 184-186, 188, 190, 193, 195, 197, 198
memory, 3, 43, 73, 74, 79, 86, 87, 156
mental plate, 193
metaphysics, 5, 16, 18, 33-36, 47, 51, 67
metaphysics of holism, 16
Michelson-Morley experiment, 24
microvita, microvitum, 59, 193
mindfulness meditation, 116

moksha, 4, 51, 152, 193, 196
monism, 5, 51, 121, 131, 138, 151
monistic idealism, 5, 147, 194
Moses, 121, 154
mudra, 120, 194
Muhammad, 127
mukti, 51, 152, 194, 196
muladhara chakra, 107-109
multiverse theory, 21, 46, 47, 163
mystic, 145, 156, 158, 194
mystical experience, 51, 74, 138, 146, 154, 155, 157, 165, 168, 194
mysticism, 146, 153

N

nadi, 108, 180, 194
Narayana, 194
natural selection, 49, 60, 61
near-death experiences, 2, 3, 101, 102, 202
Newton, 1, 7, 22, 26, 70, 79
Nicene Creed, 126
Nirguna Brahma, 52, 78, 152, 193, 194
Nonlocality, 9, 12, 16, 28, 36, 39, 40, 47, 136, 147, 163, 191, 194, 199

O

occult powers, 112, 113, 120
Ockham, Ockham's razor, 38, 40, 47, 49
Om, 54, 194
Operative Principle, 194
Oscar the cat, 44
out-of-body experiences, 3, 169

P

panentheism, 5, 148, 149, 150, 151, 152, 194, 198, 204
pantheism, 149, 151, 194
papa, 194
paradigm shift, 57, 165, 169
Paramapurusha, 52, 66, 84, 85, 92, 116, 132, 188, 189, 193-195, 198
Paramatman, 52, 85, 89, 95, 116, 144, 157, 188, 189, 194, 198, 205
Patanjali, 125
Pauli, Wolfgang, 137, 138, 203
Pauli exclusion principle, 137
Penrose, Roger, 46, 199, 200
photon, 9, 10, 11, 13, 14, 15, 28, 29, 40, 54, 69, 194, 195
Planck, Max, 8, 21, 55, 136
Planck spacecraft, 55
Planck's constant, 8, 136, 201
plasma, 55, 56
polarization, 13, 14, 15, 40
Prakriti, 52, 53, 55, 56, 59, 66-68, 73, 75-78, 84, 85, 95, 108, 151, 152, 166, 171, 189, 191, 193-198
prana, 59, 108, 194, 198
pranayama, 125, 180
prati-saincara, 57-59, 63, 65, 169, 189, 193, 194
pratyahara, 92, 117, 125
precognition, 110, 191
preconscious, 86, 182
probability, 9, 12, 21, 145
PROUT, 129, 134
psychedelic, 160, 161
psychic abilities, 40, 41, 111, 168
psychic body, 107-110
psychic power, 110, 111
psychokinesis, 110

Purusha, 52, 53, 54, 59, 67, 85, 189, 195
Purushottama, 57, 58, 59, 63, 64, 65, 122, 190, 194, 195

Q

quantum, 4, 7, 8-16, 18, 20, 28, 29, 39, 40, 47, 54, 67, 69, 84, 111, 112, 135-137, 140-145, 147, 162-167
quantum mechanics, 4, 7, 12, 13, 15, 136, 140, 164
quantum physics, 4, 7, 9, 15, 16, 28, 135-137, 142, 144, 165, 166, 195
Quran, 127

R

Radin, Dean, 3, 110, 199, 200, 202
rajoguna, 52, 53, 68, 187, 191, 195
Ramakrishna, 130, 154, 156
Ramanuja, 152
Rees, Sir Martin, 46, 200
reincarnation, 41, 83, 102, 103, 152, 195
relativity theory, 4, 31, 35, 54, 163
remote viewing, 3, 110, 188, 191, 195
Rig Veda, 119
Rinpoche, Chogyam Trungpa, 130
rishi, 77, 195
RNA, 42, 43, 72

S

Sadashiva, (see also Shiva), 51, 115, 119, 193, 195, 196, 197
sadhaka, 76, 78, 114, 118, 195
sadhana, 64-66, 81, 82, 88, 92, 97, 105, 109, 114-118, 120, 173, 176, 184,

185, 190, 193, 195
Saguna Brahma, 53, 78, 152, 194, 195
sahasrara chakra, 107-109, 125, 129
saincara, 53, 54, 57, 59, 169, 189, 194, 195
samadhi, 4, 32, 51, 81, 115, 118, 124, 125, 129, 154, 156, 157, 188, 190, 191, 195, 196
sambhuti, 120, 196
samskara, 79, 80, 93, 96, 97, 113, 196
Sanskrit, 6, 50, 52, 53, 61, 64-66, 78, 79, 81, 85, 86, 92, 94, 96, 105, 107, 108, 113-115, 117, 120, 129, 134, 152, 171, 181, 183, 186
satsang, 181, 196
sattvaguna, 52, 53, 68, 196, 191
satya, 196
Schrödinger, 140, 141, 142, 144, 146
Schrödinger Wave Function, 196
Schrödinger's cat, 140, 141
seeker, 75, 76, 114, 178
seer, 76, 77, 126, 177, 185
self, 2, 3, 6, 32, 36, 39, 41, 44, 51, 63, 72-76, 81, 85, 86, 90-92, 97, 101, 109, 112, 115, 120, 127, 128, 131, 136, 142, 151, 153, 154, 155, 158, 159, 164, 172, 174, 175, 178, 179, 182, 183, 186
self-consciousness, 142, 153, 154, 196
sentient force, 77, 196
service, 81, 97, 100, 101, 134, 136, 165, 175, 176, 180, 182, 184, 190, 192, 195
seva, 180, 183
Shaivites, 120, 196
Shakti, 196
Shankara, 127, 128
shastra, 196
Shiva, 115, 119, 120, 121, 129, 154
shloka, 196
Shunya, 124, 196

siddha, 105, 181, 196
Siddhartha Gautama, 123
siddhis, 112
sin, 54, 97, 98, 149
Sita, 83, 108
soul, 73, 93, 98, 105, 131, 152, 156, 178, 183, 186
space-time, 4, 7, 8, 13, 15, 17, 18, 22, 23, 24, 25, 26, 29, 30, 31, 32, 34, 35, 36, 39, 40, 54, 55, 75, 110, 111, 112, 167, 168, 169, 172, 187, 188, 191, 194, 196
special theory of relativity, 23
Spinoza, Baruch, 150, 151
spirituality, 6, 33, 34, 75, 122, 129, 147, 152, 153, 190, 196
St. Catherine of Siena, 154, 155
Standard Model of physics, 47, 69, 70
stars, 19, 35, 36, 44, 45, 54, 55, 56, 58, 59, 62, 70, 71, 72, 139, 157, 163, 170
static force, 59, 68, 69, 197
Steiner, Rudolf, 130
strong nuclear force, 44, 45, 68, 69, 71
strong objectivity, 1, 4
subconscious, 49, 73, 74, 86, 87, 93, 102, 104, 182, 193
Sufi, 127
Sufism, 5, 51
Sun, 24, 26, 30, 42, 55, 56, 57, 58, 70, 71, 85
superconscious mind, 50, 73, 87, 88, 168, 188, 189, 197
supernova, 24, 56, 192, 197, 201
svadhisthana chakra, 107-109
swastika, 50
synoptic Gospels, 126

T

tamoguna, 52, 53, 54, 55, 56, 59, 68, 78, 191, 197
tandava, 197
tanmatras, 63, 64, 180, 197
Tantra, 5, 51, 52, 73, 115, 120, 125, 129, 173, 188, 195, 197
Tao Te Ching, 122
Taoism, 5, 51, 122
Taraka Brahma, 197
tattva, 197
tejastattva, 55, 191, 197
telekinesis, 40, 110, 169, 191
teleological evolution, 49, 197
telepathy, 24, 40, 110, 191, 221
thermal death of the universe, 34, 35, 197
Tolle, Eckhart, 33, 130, 159, 181, 200, 206
Transcendental Entity, 119, 120, 150, 198
tsunami, 43

U

uncertainty principle, 8, 9, 111, 142
unit consciousness, 39, 59, 64, 65, 85, 94-96, 99, 105, 108, 116, 124, 125, 167, 181, 188, 189, 192, 195, 197, 198, 205
unit mind, 57, 59, 63, 66, 79, 81, 83, 85, 87, 89, 90, 94, 100, 104, 107, 108, 118, 166, 168, 169, 176, 187, 188, 190, 191, 193, 195, 198
Unity Principle, 5, 6, 15, 16, 38, 40, 49, 70, 72, 96, 98, 99, 106, 115, 116, 119-122, 124, 126, 127, 129, 130, 131, 133, 134, 136, 137, 139, 142, 146, 147, 151, 152, 154, 162, 165-172, 175, 176, 186
Upanishads, 128, 131, 142

V

Vedanta, 5, 51, 198
Vedas, 52, 196, 198
vidya, 65, 66, 75, 78, 81, 83, 92-97, 109, 110, 112, 114, 115, 124, 175, 181, 198
vijinanamaya kosha, 73, 87, 88, 108, 198
vishuddha chakra, 107, 108
Vivekananda, 130
vrittis, 196, 198

W

wave function, 9, 14, 111, 164, 166
weak force, 45, 47, 69, 70
Wheel of birth and death, 83

Y

yin-yang, 8, 123
yoga, 5, 51, 108, 117, 125, 128, 129, 130, 156, 159, 179, 180, 185, 219, 221
Yogananda, 130, 154

Z

Zazen, 182
Zen Buddhism, 182, 192

www.ingramcontent.com/pod-product-compliance
Lightning Source LLC
Chambersburg PA
CBHW021100080526
44587CB00010B/313